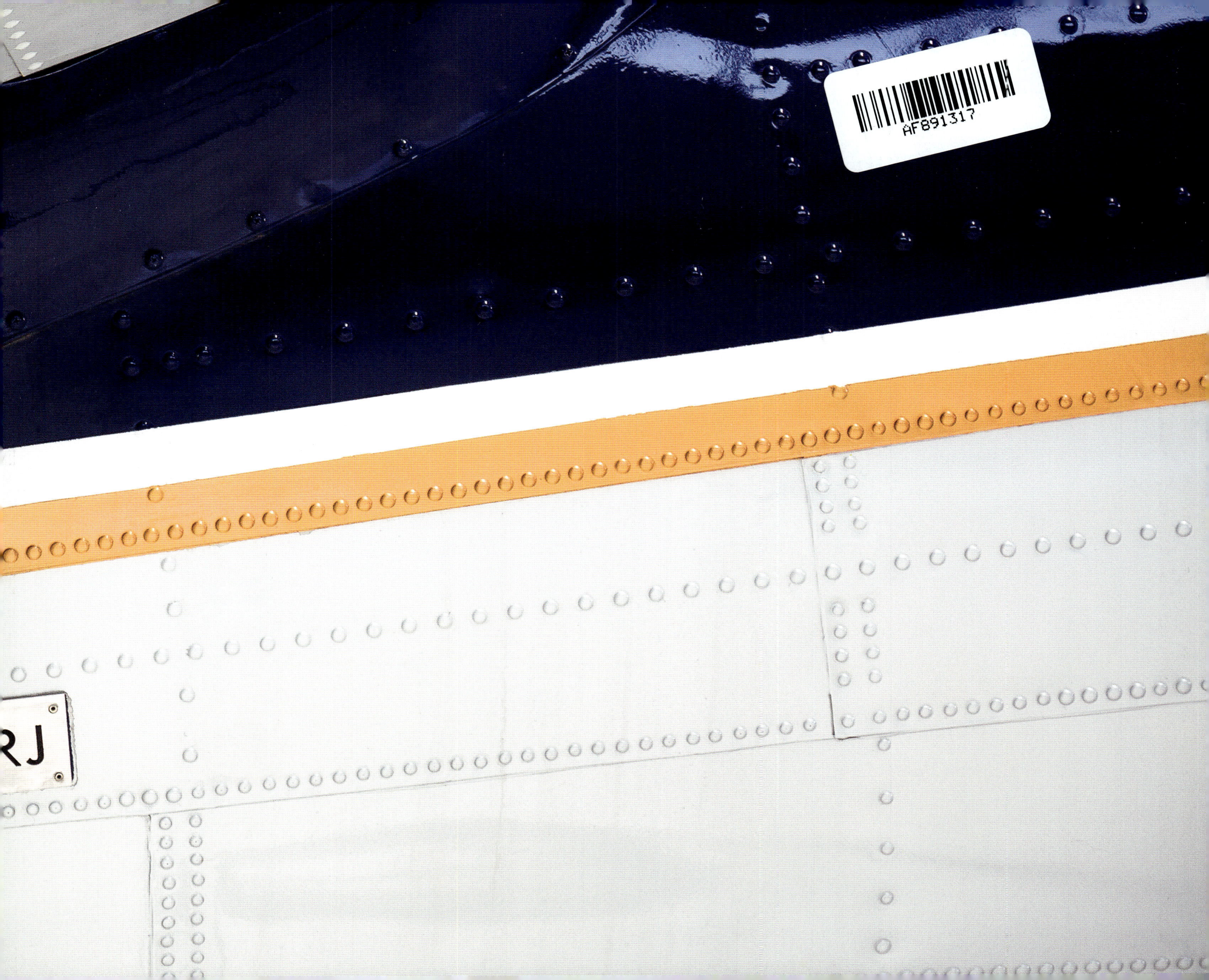
RJ

BREITLING
EXIT

DOUGLAS DC-3 BACKSTAGE

Francisco Agullo · Michael Prophet

AS Verlag

Vorwort

Ohne Zweifel: Kein anderes Flugzeug hatte je so grosse Auswirkungen auf die Geschichte der Zivilluftfahrt wie die Douglas DC-3. Sie machte das Fliegen sicher und bequem und wurde bald zum «Flaggschiff» so mancher Fluggesellschaft. Zum ersten Mal verdienten die Flugunternehmen Geld mit dem Transport von Passagieren, ohne dass sie dafür öffentliche Subventionen beanspruchen mussten.

Während des Zweiten Weltkriegs war die DC-3 in einer Militärversion als C-47 im Einsatz, zusammen mit unzähligen requirierten und an die militärischen Bedürfnisse angepassten zivilen DC-3. Die Flugzeugfabriken liefen auf Hochtouren. Sie produzierten die C-47 im unvorstellbaren Rhythmus von einem Flugzeug – pro Stunde. Am 6. Juni 1944, D-Day, dem wichtigsten Tag in der Geschichte des Zweiten Weltkriegs, verfrachtete eine Armada von C-47 Dakotas amerikanische und britische Luftlandetruppen zu Tausenden in die Normandie, an Bord Fallschirmsoldaten, im Schlepp die berühmten Gleitflieger. Nach dem Krieg wurden viele der übrig gebliebenen C-47 in Zivilversionen umgebaut. Sie wurden zur Standardausrüstung von Fluggesellschaften rund um die Welt.

Ihr Ersatz durch modernere Typen und ihre Verschrottung waren zwar nicht aufzuhalten. Aber die rundliche Form der DC-3 ziert noch heute manchen Flughafen. Bewahrte und sorgfältig gepflegte Exemplare werden leicht hundert Jahre fliegen und als erster Flugzeugtyp überhaupt seine originale Rolle über eine so lange Zeit hinweg spielen können. In der ganzen Welt vermag die DC-3 mit ihrer freundlichen Silhouette die Menschen noch heute zu berühren und lässt in ihnen lebendige Bilder der Ankunft lieber Menschen oder der Versorgung mit lebenswichtigen Gütern aufkommen.

In Anbetracht ihrer vielen Leben und ihrer fortgesetzten emotionalen Kraft könnte der DC-3 Unsterblichkeit verheissen sein. Nicht nur wegen ihrer überwältigenden Leistungen im Krieg, sondern mehr noch wegen der schieren Verlässlichkeit, die das Flugzeug bis heute auszeichnet. Viele Flugzeughersteller hatten sich den Kopf darüber zerbrochen, was für ein Flugzeug denn die DC-3 einmal ersetzen könnte – und kamen doch immer wieder auf die DC-3. Den Jungfernflug hatte das Flugzeug am 17. Dezember 1935 absolviert. Ein geschichtsträchtiges Datum: Auf den Tag genau 32 Jahre zuvor war den Gebrüdern Wright in Kitty Hawk in North Carolina der erste Motorflug gelungen. Seit 80 Jahren also ist die DC-3 immer noch irgendwo auf der Welt in Dienst. Kein anderes Flugzeug ist je auch nur in die Nähe dieser Leistung gekommen.

Als die Produktion der DC-3 und ihrer militärischen Version C-47 nach dem Krieg eingestellt wurde, waren etwa 16 000 Exemplare dieses robusten Transportflugzeugs gebaut worden, in den USA, der früheren Sowjetunion und sogar in Japan. Neben der C-47 fand die DC-3- ihren Weg in die neue Ära der zivilen Luftfahrt. Ironie des Schicksals: Statt in den 50er-Jahren langsam dem Sonnenuntergang entgegenzufliegen, flog sie weiter und überlebte alle die beflissenen Flugzeughersteller, die auf ihre Ablösung hingearbeitet hatten. Noch immer sind über hundert DC-3 im kommerziellen Transport von Passagieren und Fracht eingesetzt oder absolvieren historische Flüge. Ihre Zuverlässigkeit, die niedrige Landgeschwindigkeit und das ballonartige Fahrwerk mit dem traditionellen Heckrad ermöglichen es der DC-3 auf Dschungelpisten und unbefestigten Feldern zu landen, wo moderne Flugzeuge keine Chance hätten.

Im Jahr 2015 feiern wir den achtzigsten Geburtstag dieses grossartigen Flugzeugs. Leider verschwinden jedes Jahr mehr DC-3 und C-47 vom Himmel, dienen als Ersatzteilspender oder wandern in ein Museum. In Europa zählen wir heute noch fünfzehn DC-3 in flugfähigem Zustand. Sie werden hauptsächlich von nicht gewinnorientierten Organisationen betrieben und für historische Flüge eingesetzt. Dieser Bildband ist ein Tribut an die Menschen und Organisationen, die sich der Restaurierung des legendären Flugzeugs widmen und es in flugfähigem Zustand erhalten.

Ich danke allen Fotografen, die an der Vorbereitung zu diesem Buch mitgewirkt haben, und ebenso allen Freiwilligen, Crews und Sponsoren, die es möglich machen, dass wir die Dakota weiter am Himmel fliegen sehen.

Francisco Agullo

Editorial

The Douglas DC-3 is certainly the aircraft with the greatest impact in aviation history. It became the "flagship" of many airlines when entering service, allowing safe and comfortable transportation. For the first time, airlines were generating a profit by carrying passengers, without relying on government subsidies.

During World War ll, the military version of the DC-3 was named C-47 and many civilian DC-3s were requisitioned and converted by the army. During this period, the factories produced the C-47 at an unbelievable rhythm of one aircraft per hour.

On June 6, 1944, also known as D-Day, the most important day in the history of the Second World War in Europe, involved thousands of American and British airborne troops who arrived by parachute and glider, most dropped or towed by C-47 Dakotas. After the war, many of the remaining C-47s were converted and became the standard equipment of all airlines in the world.

The DC-3's rounded form still graces many airports while more modern types intended as its replacement have long been consigned to the scrapyard. Upgraded and preserved examples will easily see a century of flying, the first aircraft to do so in its original role. Across the world, the DC-3 affects many lives every day, its friendly shape presaging the arrival of family, equipment or vitally needed supplies.

Considering just how many lives this aircraft has, and continues to touch, its immortality is assured, not just through exploits in war as with so many other types, but through its sheer reliability in peace.

Many aircraft manufacturers had bargained for what was the best replacement for a DC-3 would turn out to be a DC-3. Auspiciously, the DC-3's maiden flight was on December 17, 1935, 32nd anniversary of the Wright brother's historic flight at Kitty Hawk, North Carolina. The plane has remained in operation, somewhere in the world, ever since and after 80 years of service. No other airliner has ever come close. When production of the DC-3 and its military derivative, the C-47 Skytrain, wound down after the war ended, some 16,000 copies of the sturdy transporter had been built in America, the former Soviet Union and even in Japan. Along with its military sibling, the DC-3 ushered in the modern era of air travel. The irony is that the DC-3, which was supposed to fly off into the sunset in the 1950s, lived on long after the aircraft-makers that sought to replace it ceased to exist. Even today, over one hundred DC-3s continue to haul freight and passengers on commercial service or on historical flights. The plane's rugged reliability, low landing speed and balloon-like landing gear with a traditional tail-wheel allow it to operate from jungle strips and muddy fields where modern planes fear to tread.

In 2015, we are celebrating the 80th anniversary of this great aircraft. Unfortunately, every year more DC-3/C-47s are disappearing from the skies, scrapped for parts or being stored in a museum. In Europe, we count today fifteen DC-3s in flying condition. They are mainly operated by non-profit organizations on historical flights.

This photo book is a tribute to those people and organizations which help restoring and maintaining this legendary airliner in flying condition.

I would like to take the opportunity to thank you all the photographers who participated in the preparation of this book and all the volunteers, crew and sponsors that allow us to see the Dakota flying in the skies.

Francisco Agullo

CELEBRATING 80 YEARS Douglas DC-3

ITY OF KANSAS CITY
TWA
U.S. MAIL
A.M.2.
GENERAL
AIR EXPRESS

Die Entstehung einer Legende

In der Bezeichnung «DC» steht das D für Douglas und C nicht etwa für Company, wie man vielleicht vermuten könnte, sondern so klar wie programmatisch für «Commercial». «Douglas Commercial».

Die DC-3 bildete den Gipfelpunkt einer Entwicklung, die mit einer Anfrage von Transcontinental and Western Airlines (TWA) an Donald Douglas ihren Anfang genommen hatte. Denn der direkte Konkurrent von TWA im transkontinentalen amerikanischen Linienverkehr, United Airlines, hatte gerade eben mit der Boeing 247 den transkontinentalen Liniendienst aufgenommen und bei Boeing eine Grossbestellung von 60 dieser Maschinen aufgegeben. Boeing lehnte es ab, irgendeine andere Fluggesellschaft ebenfalls zu beliefern, solange der United-Auftrag nicht vollständig abgewickelt sei. Also wandte sich TWA an Douglas und gab die Entwicklung eines Flugzeugs in Auftrag, mit dem das Unternehmen gegenüber United wieder konkurrenzfähig sein würde.

Daraus entstand 1933 der vielversprechende Entwurf einer DC-1, der ein Jahr später in die DC-2 mündete. Die DC-2 war ein Erfolg, hatte aber auch noch Verbesserungspotenzial.

Cyrus Rowlett Smith, Präsident von American Airlines, und William Littlewood, Vizepräsident und für die Technik zuständig, waren beide mit der DC-2 geflogen und fanden an einigen Leistungsmerkmalen des Flugzeugs wenig Gefallen, obwohl dessen Charakteristika auf dem Papier über denjenigen der Boeing 247 lagen: Die DC-2 hatte die am besten bewerteten Triebwerke, die damals in Betrieb waren, aber die beiden Herren hatten das Gefühl, die Maschine habe zu wenig Kraft. Sie beförderte 14 Personen (zwei mehr als die Vorgängerversion DC-1). Aber obwohl sie schneller war als alle anderen Flugzeuge auf dieser Route, vermochte sie die Strecke New York–Chicago nicht nonstop zurückzulegen; auch gab es Berichte, wonach das Flugzeug manchmal schwierig zu landen und die Steuerung sehr schwergängig sei. Andere Berichte über mangelnde Stabilität der Flugrichtung, über Propeller und Vereisungsprobleme beunruhigten die beiden Manager ebenso wie Behauptungen über extreme Kursabweichungen in Turbulenzen.

Smith telefonierte deshalb Donald Douglas und unterbreitete ihm einen Vorschlag: Er wollte ein besseres Flugzeug als seine bisherigen Condors oder Fords und ein besseres als die Boeing 247. Aber er wollte auch etwas Grösseres als die DC-2. Er wollte ein Flugzeug, das seiner Kundschaft Sicherheit, Komfort und Zuverlässigkeit bieten konnte. Seine Curtiss Condor «Sleepers» und die Ford Tri-Motors kamen an die von Smith gesetzten Standards einfach nicht heran.

Genesis of the Legend

The designation "DC" stands for "Douglas Commercial". The DC-3 was the culmination of a development effort that originated out of an inquiry from Transcontinental and Western Airlines (TWA) to Donald Douglas. TWA's rival in transcontinental air service, United Airlines, was inaugurating service with the Boeing 247 and Boeing refused to sell any 247s to other airlines until United's order for 60 aircraft had been filled. TWA asked Douglas to design and build an aircraft that would enable TWA to compete with United Airlines. Douglas' resulting design, the 1933 DC-1, was promising, and led to the DC-2 in 1934. While the DC-2 was a success, there was still room for improvement.

Cyrus Rowlett Smith, President of American Airlines, and William Littlewood, American Airlines Vice President of engineering, had both flown in the DC-2 and did not like some of its performance characteristics, although it was a marked improvement over the Boeing 247. It had the highest rated engines in use at the time, but they felt it lacked power. It carried 14 passengers, two more than the DC-1. Moreover, it could not make New York to Chicago, non stop, although it was faster than any other airliner on that route. They also had reports that, at times, it was difficult to land, with heavy aileron and rudder control. Additional reports of directional instability, propeller, and fin icing problems and yawing excessively in turbulence also concerned them.

Smith telephoned Donald Douglas with a proposal. Smith had decided what kind of airplane American needed. He was looking for a larger and more comfortable airplane than his Condors or Fords, and better than the Boeing 247. He also wanted something bigger than the DC-2. Smith wanted to give his customers safe, comfortable, and reliable transportation, and his Curtiss Condor "Sleepers" and Ford Tri-Motors simply did not measure up to these standards.

At first, Douglas did not react strongly or positively to Smith's proposal. He was reluctant to take on a new design and the associated headaches. The DC-2 was in full production with 102 machines already manufactured, and another 90 orders on the assembly line. A new model would mean new tooling and starting over another gamble.

Smith spent over $300 on a two-hour long distance call before he finally convinced Douglas to modify a DC-2 to American's sleeper requirements. Some have said if Smith had not persisted and made an offer, Douglas would never have built the DC-3. Douglas nevertheless was skeptical. Night flying was about as popular

TWA-Mechaniker arbeiten an der berühmten DC-2, dem Vorgängermodell der DC-3.

TWA mechanics working on the famous Douglas DC-2, the father of the DC-3.

Donald Douglas, Gründer der Douglas Aircraft Company. Er war als Ingenieur ebenso wie als Unternehmer hoch angesehen.

Donald Douglas: founder of the Douglas Aircraft Company. He was a very highly regarded engineer and a bold entrepreneur.

Donald Douglas reagierte anfänglich wenig begeistert. Es war ihm zuwider, erneut ein weiteres Entwicklungsprojekt in Angriff zu nehmen; er fürchtete die Kopfschmerzen, die ein solches Vorhaben auslösen würden. Zudem: Die DC-2 befand sich in voller Produktion, 102 Maschinen waren bereits gebaut und 90 weitere waren im Bau. Ein neues Modell hiesse: neue Pläne, neue Werkzeuge, neue Einrichtungen; das Spiel würde von vorne beginnen.

Aber Smith liess nicht locker. Er griff wieder zum Telefon, und am Ende eines zweistündigen, 300 Dollar teuren Ferngesprächs hatte er seinen Kontrahenten so weit: Douglas willigte ein, die DC-2 im Sinne der Anforderungen Smiths zu modifizieren. Kenner der Szene sollten später sagen, Douglas hätte sich nie zum Bau der DC-3 entschlossen, wenn nicht Smith so aufsässig insistiert hätte.

Dennoch: Douglas blieb skeptisch. Nachtflüge waren damals etwa so populär wie die Pest, und Douglas hegte Zweifel über Smiths Realitätssinn. Wer würde ihm wohl die nötigen Millionen vorstrecken, nur um ein Projekt zu dem Zweck zu finanzieren, damit die Leute in einem Flugzeug sollten schlafen können?

Unter den Luftfahrtunternehmen entwickelte sich jedoch ein allgemeines Einvernehmen über das Potenzial eines solchen Luftverkehrs. So begann ein detaillierter Evaluationsprozess unter Einbezug aller Bedürfnisse der Airlines, von der Zahl der Flugzeuge über den Komfort für die Passagiere und die Erfordernisse der Infrastruktur bis zu den Kosten und gesamtwirtschaftlichen Folgen und Wirkungen.

Denn die grosse Wirtschaftskrise hatte für viele amerikanische Unternehmen zu harten Zeiten geführt und die Regierung hatte durch eine eigene neue Finanzgesellschaft Mittel für den Aufbau des privaten Sektors bereitgestellt. Davon profitierte auch Smith: er erhielt für sein Projekt ein Darlehen. Am 8. Juli 1935 sandte er Douglas Aircraft ein Telegramm und bestellte zehn Flugzeuge für 795 000 Dollar. Die genauen Spezifikationen, die Smith für das neue Flugzeug verlangte, kamen bei Douglas am 14. November 1935 an, nachdem die Konstruktionsarbeiten längst begonnen hatten.

Bevor die DC-3 zum ersten Mal fliegen konnte, verdoppelte Smith seine Bestellung, sodass sie nun acht sogenannte DSTs (die Bezeichnung für die DC-3-Spezifikation Douglas Sleeper Transport) und zwölf DC-3 umfasste. Der Vertrag dazu wurde am 8. April 1936 unterzeichnet. American Airlines und Douglas waren beide eine hohe finanzielle Verpflichtung eingegangen. Heute ist es üblich, dass der Vertrag vor der Produktionsaufnahme abgeschlossen wird. Damals hatte American

as the plague, and he wondered about Smith's business sense. Where would Smith get the millions of dollars needed to finance this venture and who would want to sleep in an airplane?

After there was general agreement among the airlines on the potential, a detailed evaluation process began. The airline's total needs, from the number of aircraft, to passenger accommodations, facility requirements, and total economic impact was part of the evaluation.

The Great Depression had created hard times for many of America's industries and the government had formed the Reconstruction Finance Corporation to loan money to the private sector. Cyrus Smith took advantage of this agency and obtained a loan to fund the development of the new design. On July 8, 1935, Smith sent a telegram to Douglas ordering ten transports costing $ 795,000. The actual specifications for Smith's proposed airplane arrived at Douglas Aircraft on November 14, 1935 (long after construction had begun). Before the first flight of the DC-3, American doubled their initial order to include eight DSTs and twelve DC-3s. By the time the actual contract was signed on April 8, 1936, American Airlines and Donald Douglas both had a heavy financial commitment. In today's business environment the contract always precedes work, but in 1935, American Airlines had such faith in Douglas' dependability and integrity that the order came first and the contract after delivery.

The new aircraft was engineered by a team led by chief engineer Arthur E. Raymond over the next two years, and the prototype DST (for Douglas Sleeper Transport) first flew on December 17, 1935. In contrast to maiden flights of today's aircraft, covered extensively by the media, this flight, like the maiden flight of the DC-1 went unnoticed by the Press. But the event on a runway in Santa Monica, California, would be one of the most significant events of the twentieth century.

A version with 21 passenger seats instead of the sleeping berths of the DST was also designed and given the designation DC-3. There was no prototype DC-3; the first DC-3 built followed seven DSTs off the production line and was delivered to American.

The amenities of the DC-3 and DST (including sleeping berths on early models and an in-flight kitchen) popularized air travel in the United States. With only one or two refueling stops, eastbound transcontinental flights crossing the

Airlines aber ein derart starkes Vertrauen in die Verlässlichkeit und Integrität von Douglas, dass die Bestellung zuerst kam und erst viel später der Vertrag, als der Prototyp bereits in der Luft war.

Das neue Flugzeug wurde im Verlauf zweier Jahre von einem Team unter Chef-Ingenieur Arthur E. Raymond entwickelt. Bereits am 17. Dezember 1935 fand der Erstflug eines DST-Prototyps statt. Im Gegensatz zu Jungfernflügen heutiger Flugzeuge, über die in den Medien ausgiebig berichtet wird, blieb der DST-Erstflug wie vorher schon der DC-1-Jungfernflug von der Presse völlig unbeachtet. Trotzdem sollte sich der Start auf einer Piste in Santa Monica in Kalifornien als eines der bedeutendsten Ereignisse des 20. Jahrhunderts herausstellen.

Parallel zu dem mit Schlafkojen versehenen Typ DST wurde auch eine Version für 21 Passagiere entwickelt: Sie erhielt die Bezeichnung DC-3. Es gab aber nie einen Prototyp der DC-3. Ihr Bau schloss sich an denjenigen der DST an, nachdem davon sieben Stück produziert und an American Airlines ausgeliefert waren.

Die Annehmlichkeiten der DC-3 mit 21 Passagiersitzen und der DST mit ihren Schlafkojen und ihrer Bordküche verliehen dem Reisen mit dem Flugzeug in den USA einen grossen Popularitätsschub. Mit nur einem oder zwei Tankstopps war es jetzt möglich, die Vereinigten Staaten in etwa 15 Stunden zu durchqueren. Flüge nach Westen dauerten wegen des Gegenwinds etwa 18 Stunden, aber auch das war eine signifikante Verbesserung gegenüber der Leistung der Boeing 247. Noch wenige Jahre zuvor brauchte man in kleineren und für kürzere Strecken ausgelegten Flugzeugen einen ganzen Tag – und dazu erst noch eine Nacht in der Eisenbahn.

Im Laufe der Entwicklung der DC-3 wurden verschiedene Sternmotoren angeboten. Die frühen zivilen Exemplare waren mit Wright R-1820-Triebwerken ausgerüstet, später (und dann auch bei den militärischen Versionen) kamen Triebwerke des Typs Pratt & Whitney R-1830 Twin Wasp zum Einsatz; sie ermöglichten grössere Leistung auch in einem allfälligen Notbetrieb mit nur einem Motor.

Die frühen amerikanischen Luftfahrtunternehmen wie United, American, TWA und Eastern bestellten über 400 DC-3. Ihre Flotten bereiteten den Weg zur modernen Flugzeugindustrie. Das Flugzeug lief der bis dahin in den USA für lange Strecken bevorzugten Eisenbahn sehr bald den Rang ab.

Während des Zweiten Weltkriegs wurden viele zivile DC-3 für militärischen Gebrauch umgerüstet, und Tausende von Militärversionen der DC-3 wurden unter den Bezeichnungen C-47, C-53 und R4D gebaut. In vielen Ländern benutzten die Streit-

U.S. in approximately 15 hours became possible. Westbound trips took 18 hours due to prevailing headwinds – still a significant improvement over the competing Boeing 247. During an earlier era, such a trip would entail short hops in slower and shorter-range aircraft during the day, coupled with train travel overnight.

A variety of radial engines were available for the DC-3 throughout the course of its development. Early-production civilian aircraft used the Wright R-1820 engines, but later aircraft (and most military versions) used the Pratt & Whitney R-1830 Twin Wasp which offered better high-altitude and single engine performance.

Early American airlines like United, American, TWA, and Eastern ordered over 400 DC-3s. These fleets paved the way for the modern American air travel industry, quickly replacing trains as the favored means of long distance travel across the United States. During World War II, many civilian DC-3s were drafted

Der erste Flug einer DC-3 (eines DST-Modells) fand am 17. Dezember 1935 statt.

The first flight of the Douglas DC-3 (it was a DST model) took place on December 17, 1935.

Rechts oben: Im Jahr 1946 stürzte eine mit hochrangigen amerikanischen Militärpersonen besetzte DC-3 auf dem Gauli-Gletscher ab. Mit einer spektakulären Rettungsaktion rückte die Schweiz in den Fokus der Weltpresse. Alle Menschen an Bord überlebten.

Rechts unten: C-47 in Berlin Tempelhof in einer Reihe.
Die Dakotas wurden während der berühmten Berliner Luftbrücke (1948–1949) zum Transport von Versorgungsgütern eingesetzt.

Right above: In 1946, a Dakota with high-ranked generals and wives crashes on the Gauli Glacier. Switzerland becomes the focal point of the world press as rescue takes place. All on board survived.
Rigth below: A line of C-47s at Berlin Tempelhof Airport. The Dakotas were carrying supplies during the famous Berlin Airlift (1948–1949).

kräfte die DC-3 und ihre Militärversion als Truppentransporter, Fracht- und Sanitätsflugzeuge. Rund 16 000 Maschinen wurden produziert, einige davon auch als nicht lizenzierte Kopien in Japan (Showa LD2), andere unter Lizenz in der Sowjetunion unter der Bezeichnung Lisunov Li-2.

Nach dem Krieg wurden Tausende überschüssiger C-47 für den zivilen Dienst hergerichtet und gehörten fortan zur Standardausrüstung für die meisten Luftlinien auf der ganzen Welt. Jahrelang blieben sie im Dienst. Die breite Verfügbarkeit der Maschinen, ihr einfacher Unterhalt und ihre für die damaligen Verhältnisse beachtliche Grösse und Geschwindigkeit machten sie weltweit zum eigentlichen Treiber der Nachkriegsluftfahrt.

In den nachfolgenden Jahrzehnten wurden viele Versuche unternommen, die DC-3 zu ersetzen, einschliesslich der sehr erfolgreichen Fokker Friendship; aber kein einziger Typ kam an die Vielseitigkeit, Zuverlässigkeit und Wirtschaftlichkeit der DC-3 heran. Bis in die 70er-Jahre behielt die DC-3 eine bedeutende Rolle als Teil der Luftfahrt. Sogar noch heute, 80 Jahre nach ihrem Erstflug, ist sie bei kleinen Unternehmen noch immer im kommerziellen Einsatz. Unter Aviatik-Freaks und Piloten gilt noch immer der Spruch: «Der beste Ersatz für die DC-3 ist eine andere DC-3.»

Fluggesellschaften und die DC-3

American Airlines wurde als Mitgestalter der DC-3 über die Jahre auch ihr grösster Kunde. 114 DC-3 / DST setzte sie ein. Der Start begann mit der Auslieferung der ersten DST am 29. April 1936. Bereits am 26. Juni 1936 weihte die Gesellschaft ihren neuen Passagierdienst mit simultanen Flügen von Newark (New Jersey) und Chicago (Illinois) ein.

1936 erhielt auch die KLM ihre erste DC-3. Sie ersetzte die DC-2 im Dienst zwischen Amsterdam via Batavia (heute Jakarta) nach Sydney, der damals bei Weitem längsten fahrplanmässigen Flugverbindung überhaupt.

In Lateinamerika war Cubana de Aviación die erste Gesellschaft, welche die DC-3 einsetzte, zuerst auf Inlandverbindungen, dann aber für die erste internationale Verbindung von Havanna nach Miami im Jahr 1945.

Die DC-3 war nicht nur komfortabel und zuverlässig, sie machte das Flugtransportgeschäft auch profitabel. Für American-Chef C. R. Smith war die DC-3 das erste Flugzeug, das allein mit der Beförderung von Passagieren Geld machen konnte, und

for the war effort and thousands of military versions of the DC-3 were built under the designations C-47, C-53, R4D. The armed forces of many countries used the DC-3 and its military variants for the transport of troops, cargo, and wounded. Approximately 16,000 aircraft were produced (some as unlicensed copies in Japan as Showa L2D, and as licensed copies in the USSR as Lisunov Li-2).

After the war, thousands of surplus C-47s were converted to civil service and became the standard equipment of almost all the world's airlines, remaining in front-line service for many years. The ready availability of ex-military examples of this cheap, easily maintained aircraft (it was both large and fast by the standards of the day) jump-started the worldwide post-war air transport industry.

While aviation in pre-war Continental Europe had used the metric system, the overwhelming dominance of C-47s and other US war-surplus types cemented the use of nautical miles, knots and feet in post-war aviation throughout the world.

Numerous attempts were made to design a "DC-3 replacement" over the next three decades (including the very successful Fokker Friendship) but no single type could match the versatility, rugged reliability, and economy of the DC-3, and it remained a significant part of air transport systems well into the 1970s. Even today, 80 years after the DC-3 first flew, there are still small operators with DC-3s in revenue service. The common saying among aviation buffs and pilots is that "The only replacement for a DC-3 is another DC-3."

Airlines and the DC-3

It is fitting that as the co-creator of the DC-3 that American Airlines also be the airline that used the most machines over the years. They used 114 DC-3s / DSTs and it all started with the acceptance of the first DST on April 29, 1936. American Airlines inaugurated passenger service on June 26, 1936, with simultaneous flights from Newark, N.J. and Chicago, IL.

In 1936, KLM Royal Dutch Airlines received their first DC-3, which replaced the DC-2 in service from Amsterdam via Batavia (now Jakarta) to Sydney, by far the longest scheduled route in the world at the time.

The first airline in Latin America to use DC-3s was Cubana de Aviación, initially placing them in service on domestic routes, and then using them to start their first scheduled international service from Havana to Miami in 1945.

Oben: Swissair-Westeuropaflug für das Komitee der Landesausstellung, 22.–28. Mai 1939. Empfang in Paris-Le Bourget.
Unten: Die DC-3 HB-IRK wird hier in Bern-Belp für den Weiterflug aufgetankt und bereitgestellt. Am 18. Juni 1957 stürzte diese Maschine während eines Schulflugs in den Bodensee ab.

Above: Swissair flight to Western Europe for committee members of the Swiss National Exposition (May 22–28, 1939). The DC-3 is welcomed at Paris-Le Bourget airport.
Below: HB-IRK is fueled up at Bern-Belp airport prior to continuing its flight. The aircraft crashed into the Lake Constance during a training flight on June 18, 1957.

zwar ohne irgendwelche staatliche Subventionen. Bereits 1939 wurden 90 Prozent der Passagiere der amerikanischen Fluggesellschaften mit DC-2 und DC-3 befördert. Sowohl im zivilen wie auch im militärischen Einsatz bewies die DC-3 ihre starken Eigenschaften: robust, flexibel, leicht zu fliegen und einfach im Unterhalt. Ihre grossartigen Leistungen während des Zweiten Weltkriegs waren Stoff für Legenden.

«Wir bauten die DC-3 ohne Computer, mit dem wir sie hätten testen können. Wir hatten eine Menge Daten von der DC-1 und der DC-2, um das Design zu formulieren. So gingen wir der Sache selber auf den Grund und arbeiteten die Unterlagen aus. Wir hatten Ideen, wir hatten das Zeug, sie umzusetzen, und so liessen wir die Dinge über Nacht geschehen.» Ivar Shogran (Chief Power Plant Engineer – Douglas Aircraft Company)

Produktion

Die gesamte Produktion aller DC-3-Typen betrug 16 079 Stück. Ungefähr hundert davon sind im Jahr 2015 noch in Dienst. Im Einzelnen wurden folgende Einheiten hergestellt:

- 607 zivile Varianten der DC-3.
- 10 048 militärische C-47 und C-53 wurden in Santa Monica, Long Beach und Oklahoma City gebaut.
- 4937 Exemplare wurden unter Lizenz in der Sowjetunion gebaut. Sie trugen die Bezeichnung Lisunov Li-2 und wurden von der NATO mit dem Codenamen «Cab» versehen.
- 487 Exemplare wurden bei Mitsubishi in Showa und Nakajima, Japan, gefertigt und mit Kinsei-Triebwerken ausgerüstet (Codename der Allierten: «Tabby»).

Die Produktion ziviler DC-3 wurde 1942 eingestellt. Militärversionen wurden aber bis zum Kriegsende 1945 weiterproduziert. 1949 kam die grössere und stärkere Version einer Super DC-3 heraus. Der zivile Markt war indes noch mit C-47 überschwemmt, mit denen die Fluggesellschaften ihre Personentransport- und Frachtbedürfnisse günstig abdecken konnten. In der Folge wurden nur gerade drei Super DC-3 gebaut und 1950 für den kommerziellen Einsatz ausgeliefert. Der Prototyp der Super DC-3 stand später in der Navy unter der Bezeichnung YC-129 im Verband mit 100 auf den Stand der Super DC-3 aufgerüsteten R4D-Maschinen im Dienst.

The DC-3 was not only comfortable and reliable, it also made air transportation profitable. American's C.R. Smith said the DC-3 was the first airplane that could make money just by hauling passengers, without relying on government subsidies. As a result, by 1939, more than 90 percent of the nation's airline passengers were flying on DC-2s and DC-3s.

For both airline and military use, the DC-3 proved to be tough, flexible, and easy to operate and maintain. Its exploits during the war became the stuff of legend.

"We made the DC-3 without a computer to test it. There was plenty of data from the DC-1 and DC-2 to formulate the design. Often we got down on the floor and worked things out ourselves. There was personal ingenuity, and application, and we made things happen overnight." Ivar Shogran (Chief Power Plant Engineer – Douglas Aircraft Company)

Production

Total production of all derivatives was 16,079. Approximately 100 remain in service in 2015. Production was as follows:

- 607 civil variants of the DC-3.
- 10,048 military C-47 and C-53 derivatives were built at Santa Monica, California, Long Beach, California, and Oklahoma City.
- 4,937 were built under license in Soviet Union as the Lisunov Li-2 (NATO reporting name: "Cab").
- 487 Mitsubishi Kinsei-engined aircraft were built by Showa and Nakajima in Japan, as the L2D Type 0 transport (Allied codename "Tabby").

Production of civil DC-3s ceased in 1942; military versions were produced until the end of the war in 1945. In 1949, a larger, more powerful Super DC-3 was launched to positive reviews. However, the civilian market was flooded with second-hand C-47s, many of which were converted to passenger and cargo versions. As a result only three Super DC-3s were built and delivered for commercial use the following year. The prototype Super DC-3 served the US Navy with the designation YC-129 alongside 100 R4Ds that had been upgraded to the Super DC-3 specification.

Links: Ein Team setzt Nieten in die Cockpithülle einer C-47-Transportmaschine in der Werkstatt von North American Aviation. Die Frau links bedient einen Druckluft-Niethammer, während ihr Gegenüber den Gegenhalter führt.
Rechts: Ein Angestellter von Douglas Aircraft arbeitet am Triebwerk einer C-47. Long Beach, Kalifornien, USA, Oktober 1942.

Left: Riveting team working on the cockpit shell of a C-47 transport at the plant of North American Aviation. The woman on the left operates an air hammer, while the man on the right holds a bucking bar.
Right: Douglas Aircraft Company employee working on the port engine of a C-47 Skytrain aircraft. Long Beach, California, United States, October 1942.

Die Geschichte der BREITLING DC-3

Der Traum geht weiter

Nach dem Erfolg mit der Breitling Super Constellation nahmen wir mit zwei Freunden und unserem Sponsor Breitling ein neues Projekt auf: eines der wichtigsten Flugzeuge aller Zeiten in die Schweiz zu bringen, eine Douglas DC-3.

Dem Entschluss folgte eine ausgedehnte Suche, während der wir manche DC-3 inspizierten. Einige waren verwahrlost, andere in flugtüchtigem Zustand. Aber nur in eine «verliebten» wir uns auf der Stelle, die DC-3A N922CA der Champlain Air, einer kleinen Regionalgesellschaft in Plattsburgh N.Y.

Der Besitzer, leidenschaftlicher Liebhaber der DC-3, hatte schon mehrere solcher Flugzeuge restauriert. «Unsere» Maschine wies einen prächtigen Stammbaum auf und befand sich in ausgezeichnetem Zustand: ein ideales Flugzeug für unseren Plan, Rundflüge und Reisen anzubieten.

Nach dem Kauf mussten wir das Cockpit an europäische Standards und Vorschriften anpassen. Die Kabine verlangte ebenfalls gewisse Erneuerungen, und das ganze Flugzeug musste einer Inspektion unterzogen werden. Rumpf und Flügel wurden neu gemalt und mit dem Logo unseres Sponsors versehen. Noch vor dem Verlassen der USA mussten wir das Flugzeug in der Schweiz zertifizieren und registrieren lassen. Die gesamten Vorbereitungen nahmen etwa acht Monate in Anspruch. Dann aber, am 3. Mai 2009, waren wir bereit, die Breitling DC-3 in die Schweiz zu fliegen.

Von Miami nach Genf in 45 Flugstunden und 17 Etappen

Zwölf Passagiere und vier Crewmitglieder fanden sich zur grossen Reise ein. Das Wetter war gut, einen besseren Start hätten wir uns nicht wünschen können.

Doch schon in der ersten Stunde unseres ersten Flugs hörten wir vom Motor rechts ungewöhnliche Geräusche. Zwar zeigten alle Instrumente normale Werte an, aber wir entschlossen uns zu einer Zwischenlandung im nur wenige Meilen entfernten Sebring.

Am Boden war die Ursache rasch entdeckt: eine defekte Zündkerze. Innert einer Stunde hatten wir das Problem gelöst und konnten unsere Reise nach Ocala und Savannah fortsetzen. Am nächsten Tag landeten wir dann bei durchzogenem Wetter und lokalen Regenschauern auf dem Washington Dulles International Airport. Wir nahmen uns einen Tag frei und besuchten das fantastische Smithsonian Luft- und Raumfahrtmuseum. Das schlechte Wetter zog derweil ostwärts ab.

The History of the BREITLING DC-3

The Dream continues

Following the success we had with the Breitling Super Constellation project, we decided with two friends and our sponsor Breitling that we should bring to Switzerland one of the most important aircraft of all times: the Douglas DC-3. An extensive worldwide search followed and we inspected many DC-3s. Some were abandoned aircraft, others were in flying condition but we fell literally "in love" with Priscilla, a Douglas DC-3A registered N922CA and owned by Champlain Air, a small regional airline based in Plattsburgh, NY State.

The company owner was passionate about the DC-3 and he restored a couple of them to pristine condition. The aircraft we were interested to buy had an incredible pedigree and was in excellent condition. It was ideal for offering historical and sightseeing flights all around Switzerland and Europe.

Once the purchase of the aircraft was completed, we had to adapt the cockpit to the EASA Standards and European regulations (8.33 khz spacing, FM immunity, Mode S transponder,etc). The cabin needed to be partially redone and the aircraft fully inspected. The fuselage and wings had to be painted with our sponsor's logo and colors. Last but not least, our DC-3 had to be certified and registered in Switzerland before we left the USA. The entire process took 8 months but finally on May 3, 2009, we were ready to fly the "Breitling DC-3" to Switzerland.

From Miami to Switzerland in 17 Stops and 45 Flight Hours

On the departure date, twelve passengers and four crew members were ready for this great adventure. With a pleasant weather forecast for the first sector of this trip, we could not have any better conditions. Unfortunately, on the very first flight and with less than an hour flight time, we heard an unusual noise coming out from the right engine. In the cockpit all parameters were normal but we nevertheless decided to land in Sebring which was just a few miles away. On the ground we discovered that one spark plug was perforated. Luckily for us, in less than one hour the problem was fixed and we were able to continue our journey to Ocala and Savannah.

The next day, in poor weather and torrential rain we landed in Washington Dulles International Airport. We took a day off to visit the famous Smithsonian Aviation Museum but also to allow the bad weather to move eastward and away from our planned route.

Eine bezaubernde Flughafenangestellte posiert in Alghero auf Sardinien (Italien) vor der Breitling DC-3. September 2014.

A charming airport employee posing in front of the Breitling DC-3. Alghero, Sardinia Island, Italy. September 2014.

Links: Unsere DC-3 mit dem Namen «Priscilla», registriert als N922CA, während der Inspektion vor dem Kauf, Plattsburgh, New York, Oktober 2008.
Rechts: Ein paar Monate vor der Abreise der DC-3 in die Schweiz ist der Breitling-Schriftzug samt Logo auf dem Rumpf des Flugzeugs aufgebracht worden. Opa Locka (Miami), Februar 2009.

Left: Our DC-3 named "Priscilla" and then registered N922CA. Photo taken during the pre-buy inspection. Plattsburgh, New York, October 2008.
Right: The painters applying the Breitling decals and logo on our DC-3, a couple months before our departure to Switzerland. Opa Locka (Miami), February 2009.

3. 5. 2009 – 17. 5. 2009
1. Opa Locka (Miami) (FL, USA)
2. Sebring (FL, USA)
3. Ocala (FL, USA)
4. Savannah (GA, USA)
5. Washington (WA, USA)
6. Lewiston-Auburn (ME, USA)
7. Quebec (CAN)
8. Sept-Iles (CAN)
9. Goose Bay (CAN)
10. Wabush (CAN)
11. Kuujjuaq (CAN)
12. Iqaluit (CAN)
13. Kangerlussuaq (GL)
14. Reykjavík (IS)
15. Wick (GB)
16. Aarhus (DK)
17. Epinal (F)
18. Sion (CH)
19. Geneva (CH)

Bei gutem Wetter erreichten wir Quebec in Kanada. Dort übernachteten wir im «Château Frontenac», einem in der Nähe des St.-Lorenz-Stroms gelegenen Hotel, und genossen anderntags einen Rundgang durch die Altstadt. Aufgrund der Wetterprognosen musste ich unseren Gästen indes eröffnen, dass wir die Reise schon am nächsten Morgen fortsetzen müssten, wenn wir den Atlantik noch bei gutem Wetter überqueren wollten. Die Alternative wäre ein «Grounding» gewesen; denn eine Kaltfront mit vereisendem Regen kündigte sich an. Dafür war unsere DC-3 nicht gerüstet. Also mussten wir uns auf den Weg machen, bevor die Front hereinbrechen würde. Natürlich waren einige Passagiere über den Abbruch des Quebec-Aufenthalts enttäuscht. Aber es wurde ihnen bald klar, dass ein späterer Abflug die ganze Reise in Gefahr gebracht hätte.

Vor der Überquerung des Atlantiks nahmen wir in Goose Bay die notwenige Überlebensausrüstung auf. Die Übernachtung im weiter nördlich gelegenen Iqaluit war dann mit Temperaturen deutlich unter dem Gefrierpunkt eine Herausforderung für unser 70 Jahre altes Flugzeug. Denn ohne Vorwärmen hätten die Triebwerke beim Start leicht Schaden nehmen können. Mit der Hilfe einer lokalen Fluggesellschaft konnten wir aber den Hangar über Nacht heizen und so das Problem lösen.

Als wir am andern Tag Kangerlussuaq in Grönland erreichten, entschlossen wir uns, die Reise sofort nach dem Auftanken fortzusetzen und die lange Strecke nach Reykjavík unverzüglich in Angriff zu nehmen. In Island waren wir dann alle froh, einen Ruhetag einschalten zu können.

Wir waren jetzt in Europa. Wick in Schottland und Aarhus in Dänemark waren unsere nächsten Etappen, bevor wir schliesslich Sion und Genf erreichten, wo die Ankunft der DC-3 gefeiert wurde. Auf der letzten Strecke begleiteten uns nicht nur die Super Constellation, sondern auch Hugo Mathys' DC-3 und Beech 18.

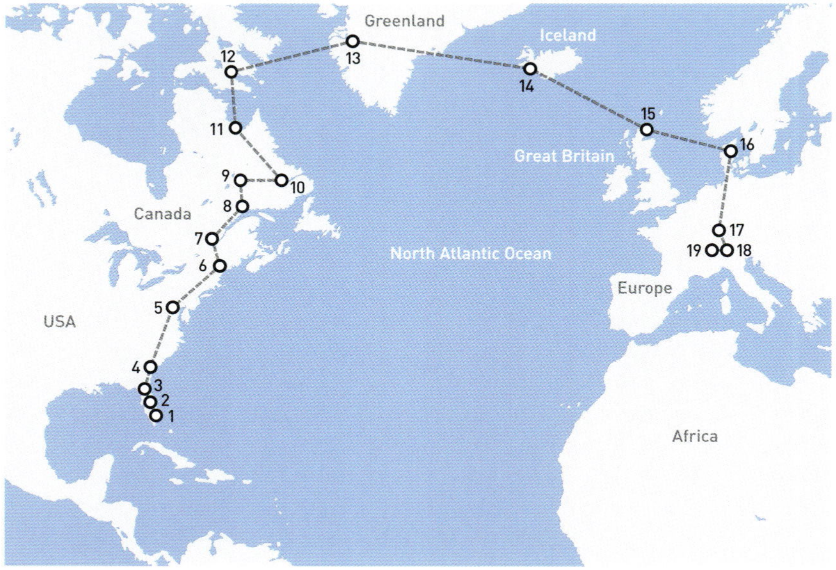

In good weather conditions we reached Canada and the city of Quebec for the night. We stayed at the Château Frontenac, a five star hotel near the St.Laurent river and all passengers were very excited to visit the old town the next day. Sadly, I had no choice but to inform them that if we wanted to cross the Atlantic in good conditions, we had to leave tomorrow morning otherwise we would be grounded as a cold front was approaching rapidly with freezing rain. Our DC-3 is not equipped to fly in icing conditions, therefore we had to be careful and had no choice but to leave before the cold front reached our area. Many passengers were disappointed as they wanted to visit the beautiful city of Quebec but they later understood that delaying the departure would have compromised the entire trip.

Before crossing the Atlantic we had to stop in Goose Bay to collect the survival equipment for the Atlantic crossing. Further north, the overnight in Iqaluit was a challenge, temperatures were well below freezing. For our 70 years old DC-3, a start in these conditions could damage the engines if no pre-heat was available. With the help of the local airline, we could park our DC-3 in the heated hangar for the night. We reached Kangerlussuaq in Greenland but decided that after refueling we would continue the long journey to Reykjavík in Iceland. We were tired and decided that we will all take a day off in Iceland.

We were now in Europe: Wick, Aarhus, and Epinal were our next stops before reaching Sion and Geneva to celebrate the arrival of the Breitling DC-3. On the last sector we were escorted by the Super Constellation together with Hugo Mathys DC-3 and Beech 18s. A great welcome committee with family and friends was waiting for us in Switzerland. It was the end of the delivery flight but the start of a new adventure.

Oben links: Damit das Öl den Motor auch bei grosser Kälte ordnungsgemäss schmieren kann, müssen die Motoren vor dem Anlassen vorgewärmt werden. Kuujjuak, Nordkanada, Mai 2009.
Oben rechts: Passagiere und Crew in ihrer «Überlebensanzügen», bereit zur Atlantiküberquerung. Goose Bay, Mai 2009.
Unten: Begrüssung in Genf mit Willkommensdusche der Feuerwehr. Genf, Mai 2009.

Above left: In cold temperatures, the radial engines of the DC-3 need to be pre-heated in order for the oil to lubricate properly the engine. Kuujjuak, North Canada, May 2009.
Above right: Passengers and crew with the "survival suit," ready for the Atlantic crossing. Goose Bay, May 2009.
Below: Welcome committee in Geneva with a "water salute" from the fire brigade trucks. Geneva, May 2009.

Links: Die Breitling DC-3 wird zum Filmstar. Nach drei Tagen Vorbereitung vor Ort ist alles bereit für das Filmteam. St. Stephan, Schweiz, Januar 2011.
Rechts: Für den Film «Gefangen im Eis» sind auf dem Rumpf Schriftfolien mit der Bezeichnung «European Air Transport Service» angebracht worden. Bulldozer platzierten Schnee rund ums Flugzeug, um den Absturz auf dem Gletscher nachzubilden.
Rechte Seite: Während der Photoflying Days konnten sich die Fotografen ganz nah zum Flugzeug begeben, um ihre Bilder zu schiessen. Zoersel, Belgien. August 2012.

Left: The Breitling DC-3 is a movie star. After 3 days of preparation on site, the team is getting ready for filming. St. Stephan, Switzerland, January 2011.
Right: New stickers "European Air Transport Service" applied on the fuselage for the movie "trapped on the ice." Trucks placed snow around the aircraft to simulate the crash on the Glacier.
Page right: During the Photoflying Days, spotters and photographers were allowed to get very close to the Breitling DC-3 for a night photo shooting. Zoersel, Belgium. August 2012.

Ein grosser Willkommensempfang mit unseren Familien und Freunden erwartete uns in der Schweiz: Das Ende des Überführungsflugs – und der Beginn eines neuen Abenteuers.

Airshows, Rundflüge und exklusive Reisen

Seit ihrer Ankunft im Jahr 2009 hat die Breitling DC-3 an vielen Airshows und Veranstaltungen für unseren Sponsor Breitling teilgenommen. Auch sind wir mehrmals im Begleitprogramm der Red Bull Air Races aufgetreten. Für unsere Mitglieder und Gäste stehen zahlreiche Optionen zur Verfügung. Wir können sie in einem einstündigen Flug rund um den Genfersee oder durch die Alpen führen. An Bord servieren wir Champagner und Feingebäck – ein hübscher Weg, die goldenen Zeiten der Luftfahrt aufleben zu lassen. Jedes Jahr organisieren wir auch eine oder zwei längere, ein- bis zweiwöchige exklusive Themenreisen. So waren wir bereits in Marokko, Spanien, Italien und Frankreich. Die Breitling DC-3 ist durch die Mitwirkung an Spiel- und Dokufilmen auch schon TV- und Kino-Star geworden.

Ihr Leben lang Passagiere befördert

Die HB-IRJ ist eine Douglas DC-3A mit der Passagiertüre auf der rechten Seite. Das Flugzeug wurde am 12. März 1940 mit der Konstruktionsnummer 2204 und dem Namen «Flagship Cleveland» an American Airlines ausgeliefert. Während des Zweiten Weltkriegs war das Flugzeug vom 4. April 1942 bis zum 1. April 1944 an die U.S. Army vermietet und in Europa stationiert. Nach dem Krieg nahm American Airlines das Flugzeug zurück und verkaufte es später, am 24. Februar 1949, an Trans Texas Airways. Diese Gesellschaft betrieb die Maschine bis 1968.

Ein Jahr später restaurierte Tradewinds das Flugzeug vollständig und verkaufte es im Dezember 1971 an die berühmte PBA (Provincetown Boston Airlines). Sie erhielt nun die Registrationsnummer N34PB und wurde von PBA bis 1987 eingesetzt. Ein Jahr später erwarb Bar Harbour das Flugzeug – und legte es vier Jahre lang still. Dann trat Champlain Air auf den Plan, kaufte die DC-3 und restaurierte sie komplett. Sie war nun als N922CA registriert.

Im September 2008 kaufte Francisco mit einer Gruppe von Freunden das Flugzeug, um es in der Schweiz fliegen zu lassen. Seit Mai 2009 wird es von der Super Constellation Flyers Association betrieben, ist in der Schweiz registriert, kann 20 Passagiere mitnehmen und ist voll instrumentenflugtauglich.

Airshows, Sightseeing Flights and Exclusive Tours

Since its arrival in 2009, the Breitling DC-3 has participated at many airshows and events for our Sponsor Breitling. We often took part in the Side Acts for the Red Bull Air Races. Another activity is to organize discovery flights. Many options are available for our members and guests, we can take them on a 60 minutes flight around the Lake Geneva or the Alps. Champagne and Petits Fours are served on board, a nice way to revive the Golden Age of Aviation. Every year we also organize longer (1–2 weeks) exclusive tours based on a theme. We have been around Morocco, Spain, Italy and France. The Breitling DC-3 has also been a cinema/TV "star" and participated in movies and documentaries.

An Entire Life Carrying Passengers

The HB-IRJ is a Douglas DC-3A with the passenger's door located on the right side of the fuselage. The aircraft was first delivered to the specifications of American Airlines on March 12, 1940, under the name "Flagship Cleveland" with the construction number 2204. In World War II, from the April 4, 1942, to the April 1, 1944, the aircraft was leased to the U.S. Army and based in Europe. Upon returning from the war, American Airlines took the aircraft back and sold it later to Trans Texas Airways on February 24, 1949. This airline which became Texas International operated the aircraft until 1968.

A year later Tradewinds restored it completely and sold it in December 1971 to the famous PBA (Provincetown Boston Airlines). Registered now as N34PB, it operated on commercial flights until 1987 when it changed colors to Eastern Express. A year later Bar Harbour Airlines leased the aircraft but also quickly put it on storage. N34PB then preserved for four years until Champlain Air bought the airplane in 1992. A complete restoration took place in 1995 and the DC-3A received a new registration N922CA. In November 2008, Francisco Agullo and a group of friends bought the airplane with the intention to make it fly in Switzerland and Europe. Since May 2009, it is operated by the Super Constellation Flyers Association. Today the aircraft is registered in Switzerland, it can carry 20 passengers and is fully IFR (Instrument Flight Rules) equipped.

Links: Eine Weltpremiere: Der erste Formationsflug eines Flugzeugs und eines Menschen. «Jetman» Yves Rossi in Formation mit der Breitling DC-3. Bei Luzern, Schweiz, Juni 2012.
Rechte Seite: Formationsflug der Breitling DC-3 und der Breitling Super Constellation. Beide Flugzeuge werden von der Super Constellation Flyers Association betrieben. Zürichsee, Schweiz, Juni 2011.
Folgende Doppelseite: Nahaufnahme der Breitling DC-3. Jedes der beiden Triebwerke entfaltet eine Leistung von 1200 PS.

Left: A world premiere: The first formation flight between an airliner and a human. Yves Rossi "Jetman" flying in formation with the Breitling DC-3. Near Lucerne, Switzerland, June 2012.
Page right: Formation Flight between the Breitling DC-3 and the Breitling Super Constellation. Both aircraft are operated by the Super Constellation Flyers Association. Lake Zurich, Switzerland, June 2011.
Following double page: Close-up photo of the Breitling DC-3. The aircraft is powered by two Pratt & Whitney R1830-92 engines. Each engine develops 1,200 horsepower.

BREITLING SUPER CONSTELLATION
BREITLING

PRATT & WHITNEY AIRCRAFT
DEPENDABLE ENGINES

Oben: Das Schilthorn ist ein 2970 Meter hoher Gipfel in den Berner Alpen mit einem Panorama-Drehrestaurant namens «Piz Gloria». Der Gipfel erlangte 1969 als Drehort des James-Bond-Films «Im Geheimdienst Ihrer Majestät» Berühmtheit.
Unten: Über dem Cockpit kann der Pilot eine Luke für den Notausstieg am Boden öffnen.

Above: The Schilthorn is a 2,970 meter high summit of the Bernese Alps in Switzerland. There is a panoramic revolving restaurant, named Piz Gloria, at the summit, which is where in 1969 the James Bond movie "On Her Majesty's Secret Service" was set.
Below: Above the cockpit, an escape hatch can be opened by the pilot and used as emergency exit for ground escape.

Die Breitling DC-3 im Flug über den Schweizer Alpen.

Hätten Sie's gewusst? *Die DC-3 kam in ihren verschiedenen Ausführungen zum Einsatz als: Luxus-Passagierflugzeug, Transportflugzeug, Frachter, Bomber, Kampfflugzeug, Transporter für Fallschirmtruppen, Schleppflugzeug für Gleiter, Feuerlöschflugzeug, Luftüberwachungsplattform, Sprühflugzeug sowie für Rundflüge und viele weitere Aufgaben.*

The Breitling DC-3 flying over the Swiss Alps.

Did you know? *The DC-3 and derivatives have been used as: luxury airliner, transport plane, cargo hauler, bomber, fighter, flying hospital, paratroopers transport, glider tow, seaplane, fire fighter, aerial survey platform, crop duster, sightseeing aircraft and many other roles.*

Die Breitling DC-3 über dem Aletschgletscher, dem grössten Gletscher der Alpen. Er ist ungefähr 23 km lang.

Hätten Sie's gewusst? *Die Douglas DC-3 ist unter einer Vielzahl von Übernamen bekannt: Doug, Dizzy, Methusalem, Gooney Bird, Skytrain, the Grand Old Lady. Aber der meistgenutzte Name ist «Dakota».*

The Breitling DC-3 Flying over the Aletsch Glacier which is the largest glacier in the Alps. It has a length of about 23 km.

Did you know? *The Douglas DC-3 is known through a variety of nicknames: The Doug, the Dizzy, Old Methuselah, the Gooney Bird, the Skytrain, the Grand old Lady, but to most of us it is the "Dakota."*

Rundflug am Matterhorn, König der Berge. Das Matterhorn ist der meistfotografierte Berg der Welt.

Sightseeing tour of the Matterhorn, the king of mountains. This famous mountain is the most photographed mountain in the world.

Links und rechte Seite: Model Laura Antonnaci posiert als Flight Attendant in der Breitling DC-3.
Folgende Doppelseite: Blick ins Cockpit: Alle Hebel und Triebwerk-anzeigen sind original; das Avionik-system der HB-IRJ hingegen ist heutiger Standard. Das Flugzeug ist für Instrumentenflug zugelassen.

Left and page right: Model Laura Antonnaci posing as Flight Attendant inside the Breitling DC-3.
Following double page: Cockpit view: while all the levers and engine instruments are original, the avionics system of HB-IRJ has been upgraded to today's standard. The aircraft is certified for IFR operations.

Hätten Sie's gewusst? *Piloten haben die DC-3 schon immer geliebt – nicht nur wegen ihrer robusten Zuverlässigkeit, sondern vor allem weil sie ohne Bordcomputer der Inbegriff für «Fliegen mit dem Hosenboden» ist.*
Schon in den 30er-Jahren wies die DC-3 zahlreiche Komfortmerk-male auf, die wir heute als selbst-verständlich betrachten, wie zum Beispiel Bordtoiletten und Bord-küchen für die Zubereitung warmer Mahlzeiten. Zum ersten Mal konnten die Passagiere in einem Flugzeug auch aufrecht stehen und sich während des Flugs frei bewegen.

Did you know? *Pilots have always loved the DC-3, not just because of its rugged reliability but because with no computers on board, it is the epitome of "Flying by seat of the pants."*
Even in the 1930s, the early Dakotas had many of the comforts we take for granted today, like on-board toilet and galley that could prepare hot food. It was the first time passen-gers were able to stand-up and walk around while the plane was airborne.

PRATT & WHITNEY
DEPENDABLE ENGINES

HB-IRJ

HYD. HAND PUMP
DOUGLAS AIRCRAFT COMPANY INC
HEATER CONTROL

Links: Regenbogen in Bern-Belp nach einem Sommergewitter.
Rechte Seite: Joe Mc Bryan, Besitzer der Buffalo Airways, beschreibt es so: «Wenn du wirklich die Erfahrung des Fliegens in diesem Leben machen willst, dann schnall' dir eine DC-3 an deinen Hintern, breite die Schwingen aus, und du bist dem Gefühl des menschlichen Flugs am nächsten.»

Left: Rainbow at Bern-Belp Airport after a summer storm.
Page right: As Joe Mc Bryan (Buffalo Airways owner) describes: "If you really want to experience flight in this life, then you have to strap a DC-3 to your ass, let the wings extend out, and that's the closest thing you will come to human flying."

Links: Einbiegen in das Rhonetal vom Genfersee her Richtung Wallis. **Rechte Seite:** Wunderschöner Blick auf die Douglas DC-3. Das Bild wurde während Air-to-Air-Aufnahmen aus der Hecktür einer Bronco geschossen.

Left: Entering the Rhône valley from the Lake Geneva towards the Valais. **Page right:** A beautiful front view of the Douglas DC-3. Photo taken from the back door of a Bronco aircraft during an air-to-air photo shooting.

Hätten Sie's gewusst? *Die Douglas DC-3/C-47 ist das erfolgreichste je gebaute Flugzeug.*

Did you know? *The Douglas DC-3/C-47 is the most successful plane ever built.*

Flug über die Normandie, 70 Jahre nach dem Zweiten Weltkrieg.

Overflying Normandy 70 years after World War II.

Hätten Sie's gewusst? *General Eisenhower zählte die Grundsteine für Amerikas Erfolg im Zweiten Weltkrieg so auf: «Der Bulldozer, der Jeep, der Halbtonnen-Laster und die Dakota.»*

Did you know? *General Eisenhower said that the foundation stones for America's success in World War II were: "The bulldozer, the jeep, the half-ton truck, and the Dakota."*

BREITLING

Links: Sechs wichtige Bedienungshebel steuern die Triebwerkeinstellungen der DC-3: Propellerhebel (2×), Gashebel (2×), Gemischhebel (2×).
Rechte Seite: Typischer Sommerflug über den Genfersee.

Left: 6 important levers control the engine parameters in the DC-3: (2×) Propeller levers, (2×) throttle levers, (2×) mixture levers.
Page right: A typical summer flight around the Lake Geneva.

Hätten Sie's gewusst? *Als Charles Lindbergh in seiner Eigenschaft als Direktor der TWA die erste Flugzeugbestellung aufgab, verlangte er von den Douglas-Ingenieuren, dass die DC-3 zu jeder Zeit auch mit nur einem Triebwerk fliegen könne.*

Did you know? *Placing the first order of the plane, Charles Lindbergh, who was a director of TWA directed to Douglas engineers that the DC-3 should always be able to fly on one engine.*

BREITLING
HB-IRJ

Sonnenuntergangsflug östlich von Cherbourg (Normandie).

Sunset flight East of Cherbourg (Normandy).

Hätten Sie's gewusst? *Der Name Dakota ist zufällig auch ein Akronym für Douglas Aircraft Company Transport Aircraft.*

Did you know? *The name Dakota, incidentally, is an acronym for Douglas Aircraft Company Transport Aircraft.*

BREITLING

Oben: Die DC-3 während einer Flugvorführung in Budapest. Diese fantastische und eindrucksvolle Schau fand während der Red Bull Air Races statt, als die Flugzeuge in geringer Höhe über die Donau flogen.
Unten: Anflug auf Piste 23 für eine tiefe Passage über das Flugfeld von Zoersel (Belgien).
Rechte Seite: Landeanflug in Sion (Wallis, Schweiz). Im Hintergrund die beiden Burgen Valère und Tourbillon.

Above: The Breitling DC-3 during a flight display in Budapest. This amazing and impressive airshow took place during the Red Bull Air Races where the aircraft are flying at low altitude over the Danube river.
Below: On final Runway 23 at Zoersel (Belgium) for a fly-by over the airfield.
Page right: On short final for landing in Sion (Valais, Switzerland). On the backgroung we can see the castles of Valère and Tourbillon.

EASTERN
THE

DC-3 Unterhalt

Ein Verkehrsflugzeug wie die DC-3 in flugfähigem Zustand zu erhalten, kann herausfordernd sein. Obwohl es in grosser Zahl hergestellt worden ist, wird es immer schwieriger, Ersatzteile zu finden. Unterhaltswerkstätten gibt es zwar noch in den USA, aber ihre Erfahrung nimmt ab. Für unsere Organisation war es von vorrangiger Bedeutung, eine Gruppe erfahrener und mit der DC-3 vertrauter Mechaniker zusammenzustellen. Aber wir brauchen auch junge Leute, um dieses Wissen zu transferieren und auf lange Sicht weiterarbeiten zu können.

In der Schweiz haben wir glücklicherweise Pius Kaufmann und Dieter Albisser. Beide arbeiteten mehrere Jahre für Classic Air, die zwischen 1986 und 2003 in der Schweiz DC-3-Flugzeuge auf kommerzieller Basis betrieb. Clive Edwards, britischer DC-3-Spezialist, gehört ebenfalls dazu. Auch arbeiten wir eng mit Paul Bazeley von Aerometal International GmbH zusammen, einer Gesellschaft in Oregon, die auf den Unterhalt von DC-3 spezialisiert ist. Wenn es nötig ist, kommt er nach Europa und unterstützt uns. Mit diesem jungen, sehr erfahrenen Team von Mechanikern und DC-3-Spezialisten sind wir in guten Händen. Unsere 75 Jahre alte DC-3 benötigt regelmässig Unterhalt und eine jährliche Inspektion. Wir wenden auch viel Zeit dafür auf, den technischen Zustand des Flugzeugs zu stärken. Auch wenn es von den Luftfahrtbehörden nicht verlangt wird, ist es doch wichtig, dass wir jedes Jahr über ein Budget verfügen, um unser Flugzeug noch besser aussehen und es noch bessere Leistungen erbringen zu lassen.

Zum vorsorglichen Unterhalt gehört es, Teile zu erneuern, die ihre Zeitlimite zwar noch nicht erreicht haben, aber doch mangelhaft werden könnten und im Falle eines Versagens zum Grounding des Flugzeugs führen könnten. Angesichts der niederen Zahl unserer jährlichen Flugstunden müssen wir sicherstellen, dass die Maschine für die Saison bereit ist und nicht von Annullationen aus technischen Gründen behindert wird. Eine Garantie dafür haben wir natürlich nicht. Aber unser Team unternimmt jede Anstrengung, um den zuverlässigen Betrieb sicherzustellen. Die Breitling DC-3 absolviert etwa 50 Flugstunden pro Jahr. Für ihren Unterhalt wenden wir im Schnitt 2000 Stunden pro Jahr auf. Das heisst: Auf eine Flugstunde kommen 40 Stunden Unterhalt.

Wenn wir unsere DC-3 kommerziell betreiben müssten, würde es nicht lange dauern, bis wir Konkurs anmelden müssten. Die Unterhaltskosten stünden in keinem Verhältnis zu den Erträgen aus dem Flugbetrieb. Die einzige Möglichkeit, sie in der Luft zu halten, ist die, dass wir uns neben unseren professionellen Mechanikern auch auf ein engagiertes Team von Freiwilligen stützen können. Die meisten von ihnen sind pensionierte Flugzeugmechaniker, einige haben einen anderen Hintergrund; aber alle teilen sie ihre Leidenschaft für die DC-3. Es gibt viel, was ein Freiwilliger leisten kann: Reinigung,

DC-3 Maintenance

Maintaining an airliner like the DC-3 in flying conditions can be a challenging task. Although the aircraft was built in large quantities, it becomes more and more difficult to find spare parts. Overhaul shops can still be found in the USA but the expertise is slowly disappearing. For our organization, it was paramount to set up a group of mechanics that have experience with the DC-3, but we also needed younger mechanics in order to transfer this knowledge and continue working in the long term.

In Switzerland, we are lucky to have Pius Kaufmann (Aerotec AG) and Dieter Albisser who worked for several years for Classic Air, a company that commercially operated DC-3s in Switzerland from 1986 til 2003. Clive Edwards, the UK DC-3 specialist is also part of our team and is always ready to help us when problems arise. We are also closely working with Paul Bazeley from Aerometal International, LLC an Oregon based company which has specialized in DC-3 maintenance, who comes to Europe to help us when needed. With this young but very experienced team of engineers and DC-3 specialists, we are now in good hands to have a reliable aircraft and a safe operation. Our 75 years old DC-3 requires regular maintenance and an annual inspection, but we also take a good amount of time each year to improve the technical condition of the aircraft. Although not required by the aviation authorities, it is important to have each year a budget to make our aircraft look and perform even better.

Known as preventive maintenance, an aircraft part might still be within limits but with time it will deteriorate rapidly and most probably a failure will occur thus grounding the aircraft. With the little amount of flying that our aircraft does each year, we need to ensure that our aircraft is ready for the flight season and that no cancellation is due to a technical reason. Of course this cannot be guaranteed but our team puts every effort to ensure we have a reliable aircraft. An average of 2,000 hours of maintenance are spent on the aircraft yearly and as the Breitling DC-3 only flies an average of 50 hours per year, you can easily deduct that for each flight hour, we need 40 hours of maintenance on the aircraft.

If our DC-3 were operated commercially, it would not take long to be in a financial bankruptcy. The maintenance cost would simply be outrageous in comparison with the revenue generated by the flying activities. The only solution for us to keep the Breitling DC-3 in flying condition is to have (next to our professional mechanics) a team of dedicated volunteers. Most of our volunteers are retired aviation mechanics others have a different background but they all share the same passion for the DC-3. There is so much that a volunteer can do: cleaning the aircraft and parts, help the mechanics, organizing logistics and parts, administration tasks, events planning and coordi-

Im Hangar von Aerometal International in Aurora (Oregon, USA) entfernt ein Mechaniker die Motorverschalung als ersten Schritt einer Triebwerkinspektion.

In the hangar of Aerometal International at Aurora (Oregon, USA), a mechanic is removing the engine cowlings, the first step in an engine inspection.

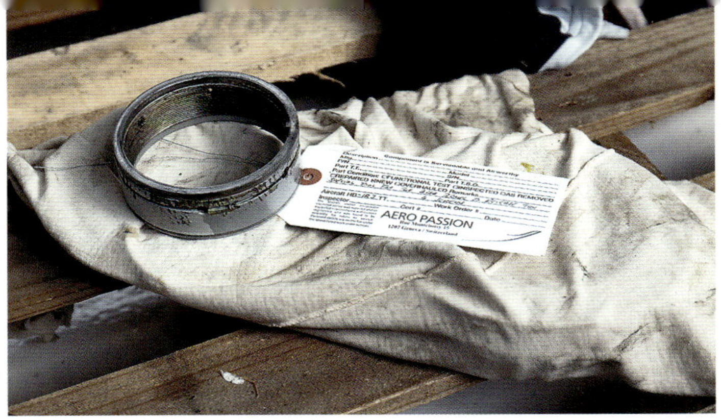

Links: Durchgehen des Unterhaltshandbuchs für die Douglas DC-3. Jeder Unterhaltsschritt ist mit dem dafür anzuwendenden verbindlichen Arbeitsverfahren präzis beschrieben.
Rechts: Jedes Teil, das aus dem Flugzeug entfernt wird, wird ordnungsgemäss etikettiert. Ein beschädigtes Teil wird entweder zerstört oder, wenn die Reparatur noch möglich ist, zur Überholung eingeschickt.
Rechte Seite: Im Hangar der Aero-Passion AG in Dole, Frankreich, ist eine grössere Inspektion im Gange. Die mittleren Bodenverschalungen sind zum Zwecke einer Korrosionskontrolle und zur Inspektion der Treibstofftanks entfernt.

Left: Going through the Douglas DC-3 Aircraft Maintenance Manual. Each maintenance task is described with the appropriate procedure to follow.
Right: Each part that is removed form the aircraft is properly tagged. The tag describes its origin, location and condition. A damaged part will either be destroyed or sent for overhaul if repair is still possible.
Page right: A major aircraft inspection is taking place at the Aero Passion hangar in Dole, France. The bottom center panels have been removed for corrosion and fuel tank inspection.

Unterstützung der Mechaniker, logistische und administrative Aufgaben, Planung und Koordination von Veranstaltungen und so weiter. Seit 2014 sind die Breitling DC-3 und die Swissair DC-3 (im Besitz von Hugo Mathys) in Dole in der Franche-Comté stationiert. Ein Hangar ist dort speziell für die DC-3, aber auch für andere historische Flugzeuge, gebaut wurden. Er kann geheizt werden, umfasst Büroräume, Küche und alle nötigen Einrichtungen; unser Team von Mechanikern und Freiwilligen arbeitet gerne dort.

Es ist jedes Mal ein Vergnügen zu sehen, mit welcher Leidenschaft unsere Mechaniker ans Werk gehen. Unermüdlich arbeiten sie am Flugzeug, sodass unsere Sponsoren, Mitglieder und Gäste ihren Flug an Bord der Breitling DC-3 geniessen können. Ohne die Hilfe dieser Mechaniker und Freiwilligen wäre unser Betrieb nicht lebensfähig. Ich kann ihnen allen nicht genug für all die Arbeit danken, die sie Jahr für Jahr erbringen.

Aerometal International – die DC-3-Doktoren

Aerometal International GmbH (AMI) wurde im Jahr 2007 mit dem Ziel gegründet, die Lufttauglichkeit älter werdender Flugzeuge aufrecht zu erhalten. Gründer und Besitzer ist Paul Bazeley. Bereits mit 16 Jahren sammelte er in der Restaurierung alter englischer Rennwagen erste Erfahrungen mit alternden Metallstrukturen. Bald folgten Flugzeuge und Sternmotoren. Inzwischen beschäftigt sich Paul Bazeley bereits seit mehr als 15 Jahren mit der Restaurierung von Flugzeugen. Schon als Kind liess er sich von der DC-3 fesseln, als er an einer Airshow Gelegenheit bekam, mit ihr zu fliegen – der erste Flug in seinem Leben.

Heute verfügt Aerometal International über breite Erfahrung in der erfolgreichen Restaurierung von Oldtimer-Flugzeugen; weltweit bekannt ist das Unternehmen aber vor allem für seine Leidenschaft und Kompetenz für die DC-3. Den Besitzern von DC-3 offeriert das Unternehmen komplette Restaurationen, den fortgesetzten Unterhalt und auch Flugoperationen. AMI betreibt in Aurora (Oregon) permanent drei DC-3 und eine C-47, ist aber mit regelmässigem Unterhalt und verschiedenen Restaurierungen von DC-3 auf der ganzen Welt engagiert.

Aerometal International deckt mit seinen Angeboten das gesamte Spektrum von Restaurierung und Unterhalt ab, einschliesslich Strukturen, Steuerung, Triebwerke, Propeller, Zubehör und weiterer Leistungen wie Bemalung, Polieren und Inneneinrichtungen. AMI nimmt umfassende Restaurationsprojekte an, ist aber besonders stolz auf seine Fähigkeit, auch detaillierte, kleine Projekte nach den Wünschen seiner Kunden abzuwickeln. Aerometal International bietet darüber hinaus auch Pilotenausbildung auf DC-3 an.

nation, etc. Since 2014, the Breitling DC-3 and the Swissair DC-3 (owned by Hugo Mathys) are now based in Dole, France. This hangar was specially built for the DC-3 but also for other historical aircraft. It is the ideal place to maintain these legendary flying machines. With a heated hangar, offices, kitchen and all the required facilities, our team of mechanics and volunteers enjoy working around the DC-3 each year.

It is always a pleasure to see how much passion and effort all our mechanics put in the maintenance. They work tirelessly on the aircraft so that our sponsors, members and guests can enjoy flying on board the Breitling DC-3. Without the help of these dedicated mechanics and volunteers, our operation would not be viable, I cannot thank them enough for all the work they do, year after year.

Aerometal International – the DC-3 Doctors

Aerometal International, LLC (AMI) was established in 2007, with the stated aim of facilitating continuing airworthiness in aging aircraft. Founder and owner Paul Bazeley has worked with antique and vintage metal structures since the age of 16, with early experience gained in the restoration of antique English racing cars. Aircraft and round engines naturally followed, and he has now worked in the vintage aviation industry for over 15 years. He first fell in thrall with DC-3s as a child when he was given a ride in one at an airshow, which happened to be his first ride in any airplane.

Today, Aerometal International has a strong record of project success with a range of vintage aircraft, but has become particularly known worldwide for its passion and capabilities related to the Douglas DC-3. Based in Oregon, USA, Aerometal International offers DC-3 owners the complete range of full restoration, continued maintenance, and flight operations of DC-3s. AMI manages three DC-3s and one C-47 full-time at its Aurora, Oregon (KUAO) facility, but takes pride in its agility and regularly performs maintenance and restoration on various DC-3s across the globe.

Aerometal International's capabilities cover the full spectrum of restoration and maintenance, including structures, flight controls, powerplants, propellers, accessories, and systems; as well as paint, polish, and interiors. AMI welcomes full scale restoration projects, but prides itself on its ability to focus on detailed and small scale projects as customers' needs dictate. Aerometal International also offers DC-3 flight training to pilots, geared to their desired outcome of either basic DC-3 flight experience, or an FAA type rating in the aircraft.

Oben: Unterhaltsarbeiten am Pratt & Whitney R1830-92-Triebwerk.
Unten: Unser Team hat soeben dieses alte und beschädigte DC-3-Höhenleitwerk aus Griechenland erhalten. Dieses Teil wird nicht mehr für unser Flugzeug verwendet, sondern in einen Konferenztisch verwandelt werden.
Rechte Seite: Kontrolle der Landeklappen und der zentralen Flügelsektion des Flugzeugs.

Above: Maintenance work on the Pratt & Whitney R1830-92.
Below: Our team just received this old and damaged DC-3 elevator from Greece. This part will not be used for our aircraft but will be transformed in a conference table!
Page right: Inspection on the flaps and center wing section of the aircraft.

Links: Inspektion des linken Fahrwerkschachts. Hier befinden sich wichtige Teile wie die Treibstoff-Boostpumpe, Feather-Pumpe und Öltank.
Rechte Seite: Unterhaltsarbeiten am Pratt & Whitney R1830-92-Triebwerk.

Left: Inspection of the left wheel well. Apart from the main landing gear components, important parts such as the fuel booster pump, the feather pump, and the oil tank can be found in this location.
Page right: Maintenance work on the Pratt & Whitney R1830-92.

EASTERN
DOUGLAS

Links: Die hintere Triebwerksverschalung wird entfernt, was den Zugang zu wichtigen Aggregaten ermöglicht wie Generator, Starter, Zündmagneten, Vakuumpumpe und Ölkühler.
Rechts: Sind die Verschalungen gereinigt, werden sie bis zum Abschluss der Flugzeuginspektion beiseitegelegt.
Rechte Seite: Letzte Anpassungen und Einstellungen am Pratt & Whitney R1830-92-Triebwerk.

Left: Removing one of the accessory cowlings. This will allow the maintenance crew to access the accessory section located behind the engine. Important components such as the generator, starter, magnetos, vacuum pump, and oil cooler are located in this section.
Right: Once the cowlings are cleaned, they will be stored until the engine inspection is completed.
Page right: Making final adjustments on the Pratt & Whitney R1830 engine.

HAMILTON STANDARD

AIR FRANCE
F-BBBE

DC-3 rund um die Welt

Es ist heute nicht mehr ganz einfach zu sagen, wie viele DC-3 rund um die Welt noch in der Luft sind. Zahlreiche DC-3/C-47 sind aufgegeben, konserviert oder für weiteren Flugbetrieb einer Restaurierung unterzogen worden. So kann man schätzen, dass weltweit noch gut hundert DC-3/C-47 fliegen. Einige von ihnen sind zur Erhöhung ihrer Leistung und Verlässlichkeit mit Turbo-Prop-Triebwerken ausgerüstet worden.

Die Dakota ist ohne Zweifel das Flugzeug mit den breitest gefächerten Einsatzformen: Es startete mit der Bezeichnung DST (für «Douglas Sleeper Transport») als Luxusliner mit Schlafkojen und avancierte schnell zum beliebtesten Flugzeug seiner Zeit. Während des Zweiten Weltkriegs wurde es intensiv für Fallschirmtruppen und als Transportflugzeug verwendet. Regelmässig beförderten die Dakota auch militärische VIPs. Vielleicht weniger bekannt ist ihr Einsatz als Löschflugzeug, in der Schädlingsbekämpfung, als Schlepper von Gleitflugzeugen oder als fliegendes Lazarett. Einige C-47 waren sogar mit Schwimmkörpern ausgerüstet, und noch heute kann man da und dort Dakotas auf Skiern entdecken.

Im Jahr 2015, 80 Jahre nach dem Erstflug der DC-3, gibt es immer noch Fluggesellschaften, die fahrplanmässige Flüge mit DC-3 ausführen. Buffalo Airways, stationiert in Yellowknife (Kanada), bietet täglich Flüge (Samstage ausgenommen) zwischen Yellowknife und Hay River (Kanada) an.

In Europa sind 15 fliegende DC-3/C-47 bekannt, und es gibt noch Projekte, zwei oder drei Maschinen zu restaurieren und für Rundflüge, Fallschirmflüge oder Demoflüge an Airshows aufzubereiten.

In den USA, hauptsächlich in Alaska und Florida, sind weiterhin C-47 für kommerzielle Frachtflüge unterwegs, und in Kolumbien ist die DC-3 noch immer ein Arbeitspferd. Ein gutes Dutzend Dakotas in mehr oder weniger flugfähigem Zustand transportieren dort lebenswichtige Güter zu weit abgelegenen Gebieten in den Llanos und im Dschungel des Amazonas. Kein anderes Flugzeug kann auf so kurzen, matschigen und unbefestigten Pisten mit Passagieren und Fracht landen. Auch in Südafrika sind noch einige DC-3, hauptsächlich für Safarireisen, im Einsatz.

In diesem Kapitel werden wir sie auf eine Reise rund um den Globus mitnehmen und ihnen fliegende DC-3/C-47 zeigen. Einige dieser Flugzeuge werden immer noch kommerziell eingesetzt und machen die DC-3 zum Flugzeug mit dem höchsten Dienstalter in der Geschichte der Luftfahrt.

DC-3 Around the World

It is certainly very difficult to say how many DC-3s are still flying today around the world. Numerous DC-3/C-47s have been derelict, preserved or are undergoing restoration for future flying activities. We can estimate that well over hundred DC-3/C-47 are still flying today worldwide. Some of these aircraft have been converted with Turbo Props to increase their reliability and improve their operational capabilities.

The Dakota is with no doubts the aircraft that has been used in the widest range of operations: it started as a luxury airliner with the DST model and its sleeping berths. It became rapidly the favorite airliner of its time. During the Second World War it was extensively used as a paratrooper and transport aircraft. In the military, Dakotas were often carrying VIPs but also hauling cargo and material. Maybe less known to the public the Dakota was also used as fire fighter, crop duster, glider tow, or a flying hospital. Some C-47s were even equipped with floats and today you can still see Dakotas on skis!

In 2015, 80 years after the first DC-3 flew there is still an airline that flies schedule flights with the DC-3. Buffalo Airways based in Yellowknife, Canada, can pride itself to be the only airline having a scheduled flight with the DC-3. Daily flights (except Saturdays) are taking place between Yellowknife and Hay River.

In Europe, we count today 15 flying DC-3/C-47s, but there are a couple of projects to restore another two or three to flying condition. Most of the European flying DC-3/C-47s are used for sightseeing flights, parachuting or displays at air shows.

In the USA, you can still find C-47s that are flying commercially for cargo operations (mainly in Alaska and Florida).

In Colombia, the DC-3 is still a workhorse. A good dozen of Dakotas (in more or less flying condition) deliver vital supplies to the most isolated communities located in the Colombian Llanos or the Amazon Jungle. No other airplane can land on short, muddy, non-prepared airstrips with such load of cargo and passengers. South Africa has also a handful of flying DC-3s, mainly used for safari tours.

In this section, we will take you around the globe, showing you some of the flying DC-3/C-47s. Some of these aircraft are still being operated commercially, making the DC-3, the airliner with the longest length of service in aviation history.

DC-3 der Air France über dem Leuchtturm von Tévennec vor der Küste der Bretagne im Département Finistère, Frankreich. Mehr Informationen über die Air France DC-3: www.francedc3.com.

The "Air France DC-3" flying over the Tévennec lighthouse located in Bretagne (Finistère), France. More information about the Air France DC-3: www.francedc3.com

Oben: Die «Swissair DC-3» im Besitz von Hugo Mathys, einem Aviatik-Enthusiasten und Flugzeugsammler.
Unten: Der Aero-Passion-Hangar in Dole, Frankreich, ist die neue Heimat für die Swissair- und Breitling DC-3 geworden.
Rechte Seite: Hugo Mathys' «Swissair DC-3» in Salzburg beim Flying Bulls-Hauptquartier. Am linken Bildrand ist eine der beiden restaurierten «Beech 18» zu erkennen; gemeinsam sind sie als «Classic Formation» bekannt und nehmen regelmässig an Airshows teil.

Above: The "Swissair DC-3" owned by Hugo Mathys, an aviation enthusiast and aircraft collector.
Below: The Aero Passion hangar (Dole, France) has become the new home for the Swissair and Breitling DC-3s.
Page right: Hugo Mathys "Swissair DC-3"and one of his two beautifully restored "Beech 18s" (known as the "Classic Formation") are regularly flying at airshows. Here in Salzburg by the Flying Bulls headquarters.

Swissair

Links: Während der Locarno Cielo Aperto Airshow 2014 hebt die «Swissair DC-3» vom Flugplatz Locarno (Schweiz) ab.
Rechte Seite: Rückflug über die Schweizer Alpen. Mehr Informationen über die Classic Formation: www.classicformation.com.

Left: The "Swissair DC-3" taking off from Locarno airport (Switzerland) during the Locarno Cielo Aperto airshow in 2014.
Page right: Return flight home, across the Swiss Alps. More information about the Classic Formation: www.classicformation.com

Swissair

Oben: Die weltweit einzige fliegende Li-2 ist bekannt als die «Russische DC-3». Von den 4924 gebauten Exemplaren ist sie die einzige heute noch flugtaugliche Maschine. Das Flugzeug ist in Budapest, Ungarn, stationiert. Mehr über die Goldtimer Foundation: www.goldtimer.hu.
Unten: Die Dutch Dakota Association ist die einzige Betreiberin in Europa mit einem AOC (Air Operator's Certificate) für die DC-3. Mehr Informationen: www.dutchdakota.nl.
Rechte Seite: Die N74598, eine C-47 A mit dem Namen «Placid Lassie» («Union Jack Dak») wurde in Long Beach (Kalifornien) gebaut und 1943 an die U.S. Army ausgeliefert. Zusammen mit anderen C-47 ihres Verbandes schleppten sie während der Invasion Waco-Gleiter (Waco für Wicks Aircraft Company) über die Küsten der Normandie. Ihre heutige Bemalung gleicht dem damaligen originalen Anstrich.

Above: The only Li-2 flying worldwide! It is known as the "Russian DC-3" and out of 4,924 examples only one remains airworthy today. The aircraft is based in Budapest, Hungary. More about the Goldtimer Foundation: www.goldtimer.hu
Below: The Dutch Dakota Association is the only operator in Europe holding an AOC (Air Operator's Certificate) for the DC-3. More information: www.dutchdakota.nl
Page right: N74598 a C-47A called "Placid Lassie" ("Union Jack Dak") was built in Long Beach (CA) and delivered to the United States Army in 1943. On early morning of June 6, 1944, she and other C-47s in her unit towed Waco gliders over the beaches of Normandy wearing the similar colors seen today.

ID
Placid Lassie
N
Eager
EILEEN
382

AIR FRANCE

Air France DC-3 vor der Aiguille du Midi (3842 m) im französischen Mont-Blanc-Massiv.

Air France DC-3 flying by the "Aiguille du Midi" (3,842 m) located in the Mont Blanc massif of the French Alps.

Oben: Fort Boyard, eine vor der Atlantikküste zwischen der Île-d'Aix und der Île d'Oléron gelegene Festung, ist der Drehort der gleichnamigen französischen TV-Spielshow.
Unten: Der Viadukt von Millau in Südfrankreich ist eine Schrägseilbrücke und überspannt das Tarntal.
Rechte Seite: Das königliche Schloss Chambord geniesst wegen seines charakteristischen französischen Renaissancestils weltweit Bekanntheit. Es liegt östlich von Blois im Département Loir-et-Cher (Frankreich).

Above: Fort Boyard: a fort located between the Île-d'Aix and the Île d'Oléron, on the west coast of France. It is the filming location for the famous TV gameshow of the same name.
Below: The Millau Viaduct is a cable-stayed bridge that spans the valley of the River Tarn near Millau in southern France.
Page right: The royal Château de Chambord at Chambord, Loir-et-Cher, France, is one of the most recognizable châteaux in the world because of its very distinctive French Renaissance architecture.

AIR FRANCE
F-BBBE

Air France DC-3 über dem Freney-Gletscher auf der Südseite des Mont Blanc.

Air France DC-3 flying over the Freney Glacier (south side of the Mont Blanc).

Oben: N18121 in der Nähe seiner Heimatbasis in Aurora, Oregon, USA.
Unten: Diese «Super DC-3» der TMS Aircraft (C-117D) kann mehr als 10 000 Pfund Zuladung transportieren. Aufgrund der FAA-Bestimmungen ist das Ladegewicht jedoch auf 7500 Pfund limitiert. Mehr Informationen: www.tmfaircraft.com.
Rechte Seite: Diese südafrikanische Springbok Classic Air DC-3 ist an einen deutschen Sammler verkauft worden. Sie ist in Deutschland und Luxemburg stationiert. Mehr Informationen: www.dakota-lodge.co.za.

Above: N18121 flying near its home in Aurora, Oregon, U.S.A.
Below: This TMF "Super DC-3" (C-117D) can carry loads in excess of 10,000 lbs but in order to comply with FAA regulations, the payload is limited to 7,500 lbs. More info: www.tmfaircraft.com
Page right: This Springbok Classic Air DC-3 from South Africa has been sold to a German collector and is now based in Europe (Germany and Luxembourg).

SPRINGBOK CLASSIC AIR

Diese Aufnahme der «Eastern Airlines DC-3» stammt von der Pazifikküste vor Oregon, nördlich der Tillamook Bay (USA).

This Photo of the "Eastern Airlines DC-3" was taken on the Pacific Coast of Oregon, just North of Tillamook Bay (USA).

Hätten Sie's gewusst? *David Edgerton, Professor für Geschichte der Wissenschaft und Technik am Imperial College in London, sagt, wir sollten uns von der Vorstellung befreien, dass die jüngste und teuerste Technologie auch immer die beste sei: «Die Tatsache, dass die DC-3 immer noch da ist und nach wie vor eine nützliche Rolle in der Welt spielt, erinnert uns auf eindrucksvolle Weise daran, dass die jüngste und teuerste Technologie nicht immer diejenige ist, welche die Geschichte verändert.»*

Did you know? *David Edgerton, professor of the history of science and technology at Imperial College, London, says we should rid our minds of the idea that the most recent inventions are always the best. "The very fact that the DC-3 is still around and performing a useful role in the world is a powerful reminder that the latest and most expensive technology is not always the one that changes history," he says.*

FLEET

Oben: Typische Schotterpiste in Alaska.
Unten: Anflug auf die Piste Whiskey-Bravo. Die kurze Piste liegt mitten in den Wäldern und wird von DC-3 zur Versorgung mit Bedarfsgütern der nahe gelegenen Kiska-Mine benutzt.
Rechte Seite: C-47 der Bush Air Cargo setzt auf Piste Whiskey-Bravo der Kiska-Mine in Alaska auf.

Above: A typical gravel airstrip in Alaska.
Below: On short final at Whiskey Bravo Airstrip. Located in the middle of the trees, this short runway is used by DC-3s to deliver supplies to the nearby Kiska Mine.
Page right: Bush Air Cargo landing C-47 on Whiskey Bravo Airstrip, Kiska Mine, Alaska.

BUSH AIR CARGO

Links: «Der einzige Ersatz für eine DC-3 ist eine andere DC-3.» Dieses von Basler Turbo Conversions of Oshkosh, Wisconsin, in eine «Turbo 67» (DC-3TP) umgebaute Flugzeug wird von zwei Turbinen Pratt & Whitney PT6A angetrieben. Die Maschine der Kenn Borek Air wird im Norden Kanadas und in der Antarktis eingesetzt.
Rechte Seite: 80 Jahre nach dem ersten DC-3-Flug betreibt die Buffalo Airways den weltweit einzigen noch existierenden fahrplanmässigen Passagier-Linienverkehr mit der DC-3. Der Dienst wird (ausser an Wochenenden) täglich zwischen Yellowknife und High River in Kanada angeboten.

Left: "The only replacement for a DC-3 is another DC-3." This Basler converted Turbo-67 (DC-3TP) is powered by two Pratt & Whitney PT6A turbines. This Kenn Borek Turbo DC-3 is being operated in North Canada and also in Antartica!
Page right: 80 years after the first DC-3 flight, Buffalo Airways operates the last scheduled DC-3 passenger service in the world. The service runs daily (except on week ends) between Yellowknife and Hay River in Canada.

Buffalo
C-GWZS

Oben: Blick aus dem Cockpit auf das Verladen von Fracht in Caruru, Kolumbien.
Unten: Im Amazonas-Regenwald ist die Versorgung abgelegener Dörfer mit Gütern des täglichen Bedarfs oft nicht möglich. Sie werden deshalb mit einer DC-3 geliefert.
Rechte Seite: Die DC-3 führt täglich Flüge in abgelegene Gegenden der Llanos und der Amazonasurwälder durch, um lebenswichtige Güter in die isolierten Gemeinden Kolumbiens zu transportieren. Diese C-47 der Air Columbia wird in Miraflores beladen.

Above: Cockpit view on the cargo loading at Caruru, Colombia.
Below: The Amazonian rain forest makes it impracticable to deliver much needed supplies to remote villages. The supplies are delivered by air on board the DC-3s.
Page right: The DC-3s make daily flights into the remote parts of the Llanos and the Amazon jungle to deliver vital supplies to Colombia's most isolated communities. This "Air Colombia C-47" is being loaded here at Miraflores.

air colombia
HK 3292
AEROPUERTO
YAMAHA
AVON
Cristar

Oben: Eine Turbo-DC-3 der Aliansa in Villavicencio. Bei diesem Flugzeug sind die originalen Sternmotoren Pratt and Whitney R1830 durch zwei Pratt & Whitney PT6A-Turbinen ersetzt worden. Solche «upgrades» werden von der Basler Turbo Conversions of Oshkosh, Wisconsin, USA, ausgeführt.
Unten: Die Grösse der Frachttüre der C-47 ermöglicht sogar den Transport eines Ersatztriebwerks R1830. Auf dem Bild wird der für eine andere DC-3 bestimmte Motor ausgeladen.
Rechte Seite: Testlauf eines Triebwerks der C-47 von Sadelca. Der weisse Rauch, der aus dem Auspuff qualmt, stammt vom Öl, das sich im Stillstand in den unteren Zylindern angesammelt hat, und beim Anlassen des Motors verbrannt wird. Der Bodenmechaniker ist mit dem Feuerlöscher bereit – für alle Fälle.

Above: An Aliansa Turbo DC-3 (Basler Turbo BT-67) at Villavicencio. On this aircraft the two original Pratt and Whitney R1830s radial engines have been replaced by two Pratt & Whitney PT6A Turbines. This "DC-3 upgrade" is carried out by Basler Turbo Conversions in Oshkosh, USA.
Below: The cargo door size of the C-47 allows to carry a spare R1830 engine inside the cabin. Here, the engine is being off-loaded and will be used on another DC-3 in need of an engine change.
Page right: Engine run on this Sadelca C-47 at La Vanguardia-Villavicencio, Colombia. The white smoke coming out from the exhaust is burned oil which accumulated in the lower cylinders. The ground mechanic is ready with its fire extinguisher... just in case.

JADELCA
POLICIA

Links: Nach einer umfassenden einjährigen Restaurierung in der Nähe von Paris wurde diese «Airscapade»-DC-3 von Frankreich nach Neuseeland überführt, wo sie nun stationiert ist.
Rechte Seite: Auf den Chathams Islands (zu Neuseeland gehörende Inselgruppe im Südpazifik) werden mit dieser prächtigen C-47 Entdeckungstouren rund um die Inseln geflogen.

Left: After an extensive yearlong restauration which took place near Paris, this Airscapade Dakota was ferried from France to New Zealand where the aircraft is now based.
Page right: In the Chatham Islands (New Zealand, South Pacific), this magnificent C-47 is operated on discovery tours around the islands.

Chathams Pacific

P
2100882
3X
Drag-em-oot

Normandie – 70 Jahre danach

Von Eugen Bürgler

Mit der beispiellosen Landung alliierter Truppen begann am 6. Juni 1944 die Befreiung Westeuropas von Nazi-Deutschland. Zum Absetzen von Luftlandetruppen spielte dabei die militärische Variante der DC-3, die C-47, eine zentrale Rolle. Zum 70. Jahrestag dieses für die Geschichte Europas entscheidenden Ereignisses trafen sich zehn DC-3 am Ärmelkanal. Mit dabei war die Breitling DC-3 mit Francisco Agullo.

Flight Attendants winkten von den Treppen moderner Businessjets und Piloten zückten ihre Handy-Kameras, als die Douglas DC-3 HB-IRJ in Breitling-Farben am 2. Juni 2014 in Genf zur Piste 05 rollte. Der elegante Propliner mit Baujahr 1940 sorgte am internationalen Flughafen zwischen all den modernen Airbus- und Boeing-Flugzeugen für Aufsehen. Nach einem 30-minütigen Hüpfer über den Jura landete die DC-3 im französischen Dole, wo sie in einem brandneuen Hangar ihre neue Heimat gefunden hat. Noch am selben Nachmittag ging es, nach einem technischen Check, in niedriger Flughöhe weiter quer über Frankreich bis zur Küste der Normandie.

Ehrung für die Veteranen

Nach gut zwei Stunden Flug erreichte die Schweizer DC-3 die französische Ärmelkanal-Küste. Die beiden Piloten, Francisco Agullo und Jon Corley, hatten die Erlaubnis erhalten, tief über jene Strände zu fliegen, wo vor 70 Jahren die Alliierten Truppen unter schweren Verlusten an Land gingen. Das satte Brummen der beiden Pratt & Whitney-R1830-92-Motoren mit je 1200 PS löste über der Normandie mehr als nostalgische Gefühle aus. Dieses Geräusch kündigte vor 70 Jahren den Beginn der Invasion der alliierten Streitmacht an. Hunderte dieser Transportflugzeuge setzten damals die ersten Soldaten – Fallschirmjäger aus den USA und Grossbritannien – ab. Die entscheidende Schlacht um Europa hatte damit begonnen, an der sich mehrere Tausend C-47 als Transporter beteiligten.

Heute fliegt die DC-3 friedlich über die Landschaft der Normandie, damals hatte die deutsche Flak aus allen Rohren auf die Flugzeuge geschossen. 70 Jahre später erinnern Gedenkstätten, Museen, die Überreste des von den Alliierten angelegten künstlichen Hafens, und vor allem die riesigen Soldatenfriedhöfe, an diese Tage. Um die Veteranen und die Gefallenen zu ehren, die damals gekämpft hatten, wurde zum 70. Jahrestag eine grosse Gedenkfeier abgehalten – in einem besonderen Rahmen, weil es wohl einer der letzten runden Jahrestage war, an dem Veteranen noch zahlreich teilnehmen konnten.

Normandy – 70 Years After

By Eugen Bürgler

On June 6, 1944, troops of the Allied Forces landed in Normandy to free Western Europe from Nazi-Germany. The C-47, a military version of the DC-3, played a key part in dropping troops by air. In commemoration of the 70th anniversary of this historic event, ten DC-3 aircraft met overhead the English Channel. Francisco Agullo's DC-3 was one of them.

Flight attendants waived from the airstairs of modern business jets, and pilots took pictures with their cellphones as the Douglas DC-3 "HB-IRJ" taxied to runway 05 at Geneva airport on June 2, 2014. The propliner which was built in 1940 was a sight for sore eyes in between all the modern Airbus and Boeing type aircraft. After a 30 minute flight across the Jura, the DC-3 touched down at France's Dole airport, where it is stationed in a brand new hangar. After some technical checks, the aircraft continued its journey at low altitude, across France, toward the coast of Normandy.

Honoring War Veterans

After approximately two hours of flying time, the Swiss DC-3 reached the French coast of the Channel. Pilots Francisco Agullo and Jon Corley were cleared to fly low overhead the beaches where Allied troops landed 70 years ago, and suffered great losses in the process. The sound of the 1200 HP Pratt & Whitney R1830-92 engines awoke emotions not only of nostalgia, but also other memories. It was that very sound which announced the beginning of the invasion by Allied Forces 70 years ago. Hundreds of these transport aircraft dropped the first soldier – American and British paratroopers – from the sky. More than one thousand C-47 aircraft were part of the deciding battle for Europe.

70 years later, the DC-3 peacefully overflew Normandy's landscapes; in the past, the German Flak would have shot at the aircraft with all available firepower. Today, memorials, museums, and the remains of a harbor set up by Allied troops, as well as military cemetaries, serve as a reminder of what happened seven decades ago. In order to honor the veterans, and the fallen soldiers who fought on D-Day, a big memorial service was held on the seventieth anniversary of the Normandy invasion. It was a very special event. This likely will have been the last event of this kind which allowed so many veterans to attend, as many of the veterans will not be around to attend the 80th anniversary of D-Day.

Die «Drag-em-oot» C-47 über der Normandie, wenige Minuten bevor Fallschirmsoldaten in Erinnerung an den D-Day vor exakt 70 Jahren aus dem Flugzeug springen werden.

"Drag-em-oot" C-47 flying over Normandy a few minutes before dropping paratroopers exactly 70 years after D-Day.

Die Breitling DC-3 nähert sich der Küste der Normandie.

The Breitling DC-3 approaching the Normandy beaches.

Hätten Sie's gewusst? Während des Kriegs wurde eine C-47 von japanischen Kampfflugzeugen angegriffen und von 3000 Schüssen in die Flügel und den Rumpf gelöchert. Das Flugzeug schaffte den Rückflug zu seiner Basis, wurde repariert und anschliessend wieder eingesetzt.
Im Jahr 1945 brach eine Dakota den Weltrekord: die Maschine flog von Pearl Harbor 1100 Meilen nach San Diego – mit nur einem Propeller, der andere Motor war defekt.

***Did you know?** During war, a C-47 was peppered with 3,000 bullets in the wings and fuselage by Japanese fighters. It made it back to base, was repaired with canvas patches and glue, and then sent back into the air.*
In 1945, a Dakota broke the world record for a flight with an engine out of action, traveling 1,100 miles from Pearl Harbor to San Diego, with just one propeller working.

Grosses DC-3-Treffen

Nachdem die DC-3 ihre Runden über der Normandie gedreht hatte, nahm sie Kurs auf den Ärmelkanal hinaus und steuerte den Flugplatz Lee-on-the-Solent bei Southampton an. Am D-Day, wie der 6. Juni 1944 auch genannt wird (Decision oder Debarkation Day), war Lee-on-the-Solent einer der wichtigsten Flugplätze und diente in erster Linie als Basis für Spitfire- und Mustang-Staffeln, die alleine am 6. Juni zu 435 Einsätzen über der Normandie starteten.

Im Anflug auf Lee-on-the-Solent waren bereits drei C-47 auf dem Flugplatz auszumachen, zu denen sich die Schweizer DC-3 gesellte. Insgesamt acht der legendären Transportmaschinen hatten sich auf dem grosszügigen Flugplatzgelände eingefunden, nachdem am 3. Juni vier weitere Exemplare gelandet waren. Sie gehören unterschiedlichen Versionen der DC-3-Familie an, am häufigsten vertreten waren die militärischen Versionen C-47 und C-53, dazu kam eine Lisunov Li-2, die sowjetische Lizenzvariante der DC-3, und mit der HB-IRJ, die einzige originale Douglas DC-3 – eine der ganz wenigen überhaupt, die weltweit noch fliegen.

Fallschirm-Absprünge aus DC-3

Das Round Canopy Parachuting Team war Initiant dieses Treffens. Diese Vereinigung will die grösste Invasion der Geschichte nicht in Vergessenheit geraten lassen und führt zu Ehren der alliierten Soldaten Fallschirmabsprünge mit Rundkappenschirmen durch, die den damals verwendeten Schirmen ähnlich sind. Zum 70. Jahrestag des D-Day hat sich das Team das Ziel gesetzt, möglichst viele DC-3 zurück in die Normandie zu holen und an historischen Stätten im Rahmen der Gedenkfeiern Fallschirmabsprünge durchzuführen.

Alle Oldtimer waren hervorragend gepflegt. Ausnahmslos haben sie beeindruckende Lebensläufe hinter sich. Dass sie noch lufttüchtig sind, zeugt von den unglaublichen Qualitäten der DC-3. So fliegt zum Beispiel die 1942 gebaute C-47 heute noch als N74589. Sie war am D-Day 1944 in den Farben, die sie heute wieder trägt, Teil der Invasionsflotte.

Big DC-3 Gathering

After the DC-3 completed its round overhead Normandy's beaches, it flew across the English Channel, and landed at Lee-on-the-Solent, close to Southampton, England. Lee-on-the-Solent was of great importance on D-Day (Decision or Debarkation Day). It served as a base for the Spitfire and Mustang fleet which departed on 435 missions over Normandy on June 6 alone.

The Swiss DC-3 joined three C-47 aircraft which were already parked on the tarmac. A total of eight of these legendary transport aircraft gathered at the airfield, after four additional aircraft landed at the airport on June 3. They are all different versions of the DC-3 family; most of which were C-47's, and C-53's. There was also a Lisunov Li-2, which is the Soviet version of the DC-3, as well as HB-IRJ, the only original Douglas DC-3 aircraft, one of the very few which are still airborne worldwide today.

Parachute Jumps from the DC-3

The gathering was initiated by the Round Canopy Parachuting Team, an organization dedicated to the memory of the greatest invasion in history. In honor of Allied Soldiers, the team conducts parachute jumps with round canopies resembling those used on D-Day. In commemoration of the 70th anniversary of the Normandy invasion, the team tried to gather as many DC-3 type aircraft as possible overhead Normandy's beaches, and to conduct parachute drops at historic sites as part of memorial services.

All vintage aircaft were very well maintainted, despite the long, impressive history that some of them may have had. The fact that so many of them are still in airworthy condition is an impressive testament to the qualities of the DC-3. A C-47 which was built in 1942, for example, still flies today, registered as N74589; that same aircraft was part of the fleet which invaded Normandy on D-Day. Its paint scheme has remained the same.

Oben links: D-Day-Atmosphäre.
Oben rechts: Das Round Canopy Parachuting Team in Originaluniformen von 1940 macht sich für den Flug über den Ärmelkanal bereit. Lee-on-the-Solent, UK. 6. Juni 2014.
Unten: Gruppenbild der Fallschirmjäger in Lee-on-the-Solent, UK. Im Hintergrund die C-47 «Union Jack Dak», die am D-Day teilgenommen hatte.

Above left: D-Day atmosphere.
Above right: In their original uniforms from the 1940s, the Round Canopy Parachuting Team is getting ready for the flight across the Channel. Lee-on-the-Solent, UK. June 6, 2014.
Below: Paratroopers group photo at Lee-on-the-Solent, UK. The C-47 "Union Jack Dak," which participated at the D-Day, can be seen in the background.

Oben: Fallschirmjäger machen sich bereit, an Bord der C-47 zu steigen. Cherbourg, Normandie.
Rechts: Insgesamt acht Dakotas (C47, C53, DC-3 und Li-2) waren zum Start Richtung Normandie zu den Feiern des 70. Jahrestages des D-Day in Lee-on-the-Solent versammelt.

Above: Paratroopers getting ready to board a C-47 at Cherbourg, Normandy.
Right: A total of eight Dakotas (C47s, C-53s, DC-3 and Li-2) were together at Lee-on-the-Solent to celebrate the departure to Normandy and the beginning of the D-Day celebrations.

AIRLINES
1942-2012

Oben: Handschlag mit Peter Braun, Mitbegründer des Round Canopy Parachuting Team. Peter war selber Jumpmaster (Absetzer) auf C-47 Dakotas.
Unten: Fallschirmjäger mit ihren Rundkappenfallschirmen bereit zum Absprung per Aufziehleine aus der C-47.
Rechte Seite: Dänische Dakota-Freunde mit C-47A in Formation mit dem finnischen Veteran C-53C über dem Ärmelkanal.
Folgende Doppelseite: Die Breitling DC-3 HB-IRJ, genau 70 Jahre nach dem D-Day, in der Nähe des Utah Beach in der Normandie, am 6. Juni 2014.

Above: Handshake with Peter Braun, Co-founder of the Round Canopy Parachuting Team. Peter was also acting as Chief Jumpmaster in the C-47s.
Below: Paratroopers with their round canopy parachutes performing static line jumps from the C-47.
Page right: Danish Dakota friends C-47A in formation flight across the Channel with the Finnish Air Veteran C-53C.
Following double page: The Breitling DC-3 HB-IRJ, exactly 70 years after D-day, flying near Utah Beach in Normandy on June 6, 2014.

AERO
FINNISH AIRLINES
K-682
FORENINGEN FOR FLYVENDE MUSEUMSFLY
OY-BPB

BREITLING
HB-IRJ

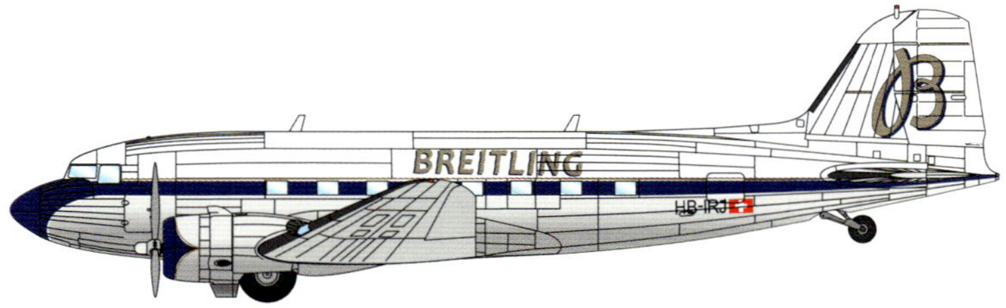

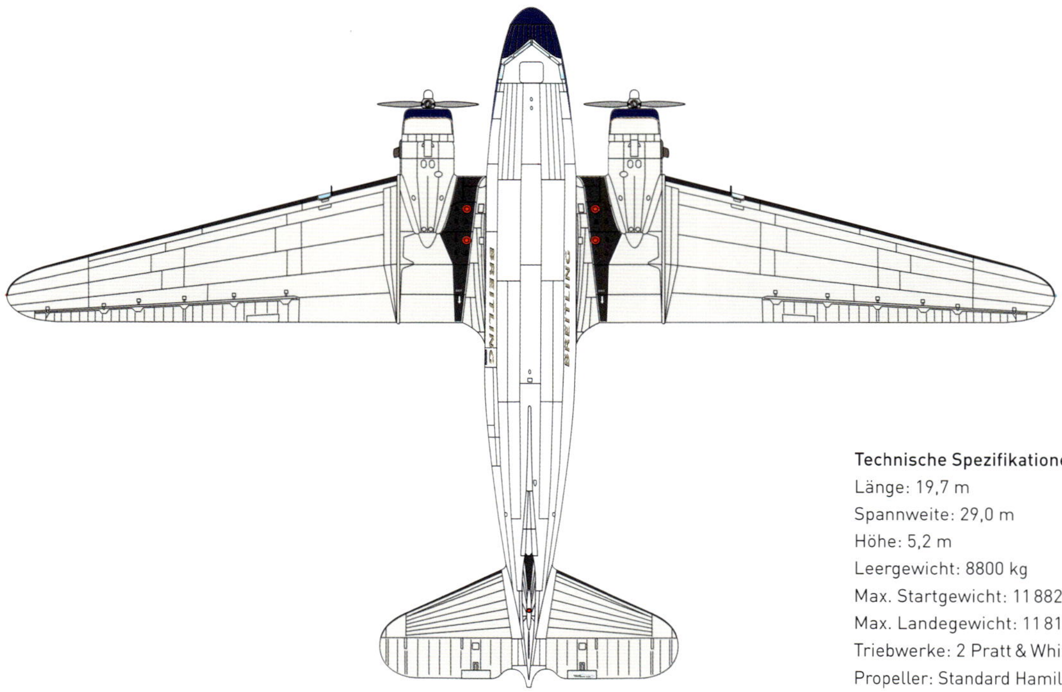

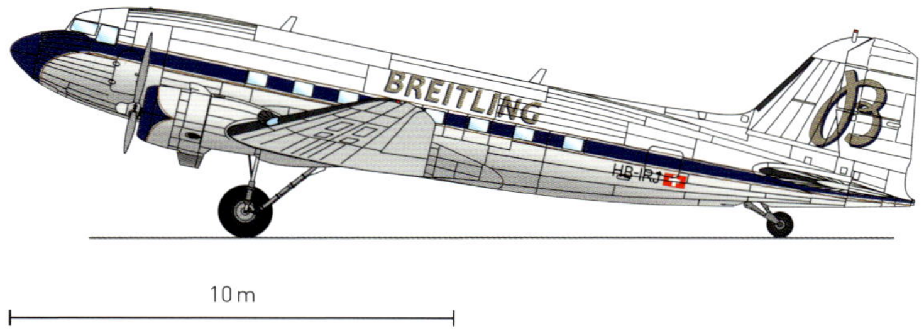

Technische Spezifikationen der Breitling DC-3
Länge: 19,7 m
Spannweite: 29,0 m
Höhe: 5,2 m
Leergewicht: 8800 kg
Max. Startgewicht: 11 882 kg
Max. Landegewicht: 11 818 kg
Triebwerke: 2 Pratt & Whitney R1830-92 mit je1200 PS
Propeller: Standard Hamilton, variable Propellerstellung bei konstanter Drehzahl
Anzahl Sitze in der Kabine: 21–30
Crew: 3
Reisegeschwindigkeit: 250 Knoten
Startgeschwindigkeit: 85 Knoten
Anfluggeschwindigkeit: 85 Knoten
Stall-Speed VSO (Abrissgeschwindigkeit in Landekonfiguration): 64 Knoten
Benötigte Pistenlänge: 900 m
Maximale Flughöhe: 18 000 Fuss
Treibstoffverbrauch: 380 l/h
Ölverbrauch: 6 l/h
Tankkapazität: 3050 l
Reichweite: 8 Stunden oder 1080 nautische Meilen (ohne Reserve)
Erstflug der DC-3: 17. Dezember 1935

Specifications Breitling DC-3
Length: 19,7 meters
Wingspan: 29,0 meters
Height: 5,2 meters
Empty weight: 8,800 kg
Maximum take-off weight: 11,882 kg
Maximum landing weight: 11,818 kg
Engines: Pratt & Whitney R1830-92, horsepower 1,200 each
Propeller: Standard Hamilton variable pitch constant speed
Number of seats in the cabin: 21–30
Crew members: 3
Cruising speed: 135 knots
Take-off speed: 85 knots
Approach speed: 85 knots
Stall speed VSO: 64 knots
Runway length necessary: 900 meters
Ceiling: 18,000 feet
Fuel consumption: 380 liters/hour
Oil consumption: 6 liters/hour
Fuel capacity: 3,050 liters
Range: 8 hours or 1,080 NM (without reserve)
Maiden flight of the DC-3: December 17, 1935

Zeichnungen / Drawings: © 2015 Terry Lawless

DC-3 Cockpits

Während der 80 Jahren, in denen DC-3 inzwischen in Betrieb sind, haben sich Technik und Technologie weiterentwickelt – und so auch das Cockpit der DC-3. Viele DC-3 sind mit moderner Avionik aufgerüstet worden, andere haben die ursprünglichen Einrichtungen behalten.

After 80 years of DC-3 operations, technology has evolved and DC-3 cockpits as well. Many DC-3s have been upgraded with new avionics while others have kept the basic equipment.

1 DC-3A, HB-IRJ, «Breitling DC-3», Aero Passion S.A. / Super Constellation Flyers Association, Switzerland

2 DC-3C, N431HM, «Swissair DC-3», Hugo Mathys, Switzerland

3 Li-2, HA-LIX, «Malev Li-2», Goldtimer Foundation, Hungary

4 DC-3-65TP, N200MF, Missionary Flights International, USA

5 C-47A, TC-ALI, Transport Museum Istanbul, Turkey

6 C-47A, N473DC, Dakota Heritage, UK

7 DC-3-65TP, N146RD, Lee County Mosquito Control District, FL, USA

8 C-47A, CF-FTR, FNT Inc, Canada

9 C-47A, N47E, «Miss Virginia», USA

10 C-47A, N345AB, «Whiskey 7», National Warplane Museum, USA

1

2

3

4

5

6

7

8

9

10

DC-3 Interiors

Die DC-3 ist für die unterschiedlichsten Zwecke eingesetzt worden. So sind auch die Inneneinrichtungen an die spezifischen Missionen angepasst: Fracht, Passagiere, Fallschirmtruppen oder VIP.

The DC-3 has been used in many different type of operations. The aircraft interiors are adapted to each specific mission, being to carry cargo, passengers, paratroopers, or VIPs.

1 DC-2, PH-AJU, KLM Colors, Aviodrome, Netherlands.
2 DC-3C, N431HM, «Swissair DC-3», Hugo Mathys, Switzerland.
3 DC-3C, N877MG, «Pan Am DC-3», Historic Flight Foundation, USA
4 C-117D, CP-2421, Lineas Aereas Canedo, Bolivia
5 C47B, 9Q-CUK, Vallentuna Aviatörförening, Sweden
6 C-47A, N345AB, «Whiskey 7», National Warplane Museum, USA
7 Li-2, HA-LIX, «Malev Li-2», Goldtimer Foundation, Hungary
8 C-47A, N47E, «Miss Virginia», USA
9 C-47A, N74589, «Union Jack Dak», England
10 C-117D, N30TN, TransNorthern Aviation, Alaska, USA.

1

2

3

4

5

6

7

8

9

10

Das Autorenteam

Francisco Agullo. Die Leidenschaft für die Fliegerei hatte Francisco Agullo schon als Kind gepackt. Als er sein Ingenieur-Diplom in der Tasche hatte, entschied er sich dafür Buschpilot zu werden. Damit verdiente er sich seinen Lebensunterhalt in Nordamerika, später in Afrika. Als er nach Europa zurückkehrte, wurde er Linienpilot und flog Maschinen von Douglas und Boeing. Im Jahr 2000 gründete er zusammen mit Freunden die Super Constellation Flyers Association, welche eine der beiden weltweit einzigen flugfähig gebliebenen Lockheed Super Constellations betreibt. Heute kümmert sich Francisco Agullo um seine wunderbar restaurierte DC-3. Francisco Agullo hat über 12000 Flugstunden auf 45 verschiedenen Flugzeugtypen absolviert. Er ist heute Flugkapitän auf privaten Jets sowie Fluginstruktor und Prüfungsexperte auf der DC-3 und auf der Super Constellation.

Michael Prophet. 1956 geboren auf der Karibikinsel Curaçao der damaligen Niederländischen Antillen. Schon in jungem Alter war Michael S. Prophet von der Luftfahrt in allen ihren Erscheinungsformen fasziniert. Als er fünf war, zog seine Familie in die Niederlande. Den entscheidenden Kick bekam seine Liebe zur Luftfahrt, als er 1974 sein erstes Transatlantik-Ticket kaufte und mit Pan Am in einer Boeing 707 nach New York flog, um dort Freunde zu besuchen. Dies löste auch das Interesse am Fotografieren aus. Ein grosser Teil seines Lebens ist seit jener Zeit der Luftfahrt gewidmet. Nach Abschluss der Oberschule begann er sogleich bei KLM als Junior-Flugzeugmechaniker, wo er an DC-8 sowie an den Grossraumflugzeugen B747 und DC-10 arbeitete. In den frühen 80er-Jahren fielen ihm erstmals die «Osprey Sky Truck»-Bücher in die Hände. Ihr Autor, Stephen Piercey, wurde mit seinem einzigartigen Stil und Humor so etwas wie ein Mentor und Inspirator für Michael. Das war der Beginn einer lebenslangen Liebe für die schrumpfende Zahl klassischer Propellerflugzeuge rund um die Welt.

The Authors

Francisco Agullo. His passion for aviation started when he was a kid. After completing an engineering degree he decided to become a bush pilot, initially in North Canada and a few years later in Africa. After returning to Europe, he becomes an Airline Pilot, flying Douglas and Boeing Airliners. In 2000, he creates with his friends the Super Constellation Flyers Association which operates today one of the last two airworthy Lockheed Super Constellation in the world. Today he looks after the beautifully restored Breitling DC-3. Francisco has accumulated over 12,000 flight hours in 45 different aircraft types. He is today Captain on private jets and Flight Instructor/Examiner on the DC-3 and Super Constellation.

Michael Prophet. Born 1956 on the Island of Curaçao, Dutch Antilles. From a young age, Michael S. Prophet became fascinated with aviation in all its aspects. After moving to The Netherlands, his love for aviation really kicked off when he purchased his first transatlantic ticket with Pan Am, flying on a Boeing 707 jetliner to visit some friends in New York, back in 1974. This also triggered the interest in taking photographs. Much of his life since then, has been devoted to aviation. Just out of high school, he started working at KLM Royal Dutch Airlines as a Junior Aircraft Mechanic, working on the classic DC-8 and B747/DC-10 wide-body jets. In the early eighties he first got hold of the famous *Osprey Sky Truck* photo books. It's author, Stephen Piercey, with his own unique style and humor, became like a mentor and inspiration to him. This was the moment that he started a life-long love for the dwindling number of classic prop-liners around the globe.

www.as-verlag.ch

Gestaltung und Herstellung / Design and Production: AS Verlag, Urs Bolz, Zürich
Übersetzung ins Deutsche: Rudolf Messerli, Oberwil BL
Translation in English: Catherine Frei, Merrimack/NH, USA
Korrektorat / Proofreading: Ellen Elfriede Schneider, Pratteln
Druck / Print: B & K Offsetdruck GmbH, Ottersweier
Einband / Cover: Grossbuchbinderei Josef Spinner GmbH, Ottersweier
ISBN: 978-3-906055-26-8

Bildnachweis · Picture Credits

Gregory le Moigne: Umschlag vorne / Front cover, 26, 27, 28, 29
Urs Bolz: 2/3, 48, 49, 50, 51, 52, 58
Gabor Hajdufi: 4, 21
TWA Museum Kansas City (Michael Prophet collection): 8
Boeing archives: 10, 11
AS Verlag archives: 13 (oben links / top left)
U.S. Navy National Museum of Naval Aviation:
13 (unten links / bottom left)
Daniel Ruhier collection: 13 (oben und unten rechts / top and bottom right)
Wikimedia Commons, United States Library of Congress,
Author T. Palmer: 15
Luigi Canu: 16
Francisco Agullo collection: 18, 19, 20
Katsuhiko Tokunaga: 22, 23
Chris Haag: 24/25
Peter Kleynen (Erci Coeckelberghs collection): 30, 31
Lyle Jansma: 32/33, 35, 38/39, 42/43, 46, 53, 54, 55,
70 (oben / top), 72/73, 84, 86, 87 (oben rechts und unten / top right and bottom), 88, 90, 91, 92/93
Unkown: 34
Skynews (Hansjörg Bürgi): 36, 37
Alexandra le Blanc (Images Essentially): 40
Marie-Dominique Schenk: 41
Hans Scholtens: 44 (unten / bottom)
Vincent Jendly: 45
Gilles Bordes-Pagès: 56, 64/65, 66, 67, 68/69
Michael Prophet: 59, 62, 63, 70 (unten / bottom),
71, 80 (oben / top), 96, 97, 98, 99
Erich Gandet: 60, 61
Bernd Sturm: 74, 75, 78, 79, 81
Wikimedia Commons: 76
Buffalo Airways: 77
Peter Gralla: 80 (unten / bottom)
Airscapade collection: 82
Brendan Odell: 83
Eugen Bürgler: 87 (oben links / top left)
Richard Davies: 89

Dank

Mein herzlicher Dank geht an:
- AS Verlag, der die Publikation dieses DC-3-Bildbands ermöglicht hat;
- Matthias Weber und Urs Bolz für die grossartige Zusammenarbeit in der Aufbereitung dieses Buchs;
- Michael Prophet für seine Unterstützung in den Recherchen und die grossartigen Bilder;
- Lyle Jansma für die riesige Arbeit und die hohe Qualität ihrer Bilder;
- Rudolf Messerli für die Übersetzung ins Deutsche;
- Catherine Frei für die Übersetzung ins Englische;
- Ellen Elfriede Schneider für das Korrektorat;
- alle Fotografinnen und Fotografen, die mir das Abdruckrecht für die in diesem Buch publizierten Bilder überlassen haben;
- alle, die mich bei diesem Projekt unterstützt haben.

Francisco Agullo

Thanks

I would like to thank:
- AS Verlag to allowing me the publication of this DC-3 photo book;
- Matthias Weber and Urs Bolz for the great cooperation in the preparation of this book;
- Michael Prophet for his support in the research of information and the great pictures;
- Lyle Jansma for the tremendous work accomplished and the high quality of the photos taken;
- Rudolf Messerli for the translation work in German;
- Catherine Frei for the translation work in english;
- Ellen Elfriede Schneider for the proof reading of my texts;
- all photographers who have given me their permission to have their pictures published in this book;
- all who have morally supported me during this project;

Francisco Agullo

BREITLING
1884
SWISS
MADE
MPH
BASE
1000M